Green Energy and Technology

Climate change, environmental impact and the limited natural resources urge scientific research and novel technical solutions. The monograph series Green Energy and Technology serves as a publishing platform for scientific and technological approaches to "green"—i.e. environmentally friendly and sustainable—technologies. While a focus lies on energy and power supply, it also covers "green" solutions in industrial engineering and engineering design. Green Energy and Technology addresses researchers, advanced students, technical consultants as well as decision makers in industries and politics. Hence, the level of presentation spans from instructional to highly technical.

Indexed in Scopus.

Indexed in Ei Compendex.

Benjamin Duraković

Zero Energy Buildings

Fundamentals, Materials and Technologies for a Sustainable Future

Springer

Benjamin Duraković
Faculty of Engineering and Natural Sciences
International University of Sarajevo
Sarajevo, Bosnia and Herzegovina

ISSN 1865-3529 ISSN 1865-3537 (electronic)
Green Energy and Technology
ISBN 978-3-032-16026-3 ISBN 978-3-032-16027-0 (eBook)
https://doi.org/10.1007/978-3-032-16027-0

This Springer imprint is published by the registered company Springer Nature Switzerland AG
The registered company address is: Gewerbestrasse 11, 6330 Cham, Switzerland

Preface

The global energy landscape is undergoing a profound transformation due to the need for sustainable energy use and mitigation of greenhouse gas emissions. In this transformation buildings offer significant potential in reducing these emissions through the energy efficiency strategies and renewable integration.

This book, *Zero Energy Buildings: Fundamentals, Materials and Technologies for a Sustainable Future*, is attempt to provide a comprehensive guide to the principles, strategies, and technologies that enable the design, construction, and retrofitting of energy-efficient buildings. The book discusses the fundamental concepts and strategies of zero energy buildings (ZEBs), as well as innovations in construction materials and technologies used to achieve energy-efficient buildings. The chapters are intentionally structured to guide the reader from a broad understanding of global energy challenges and conceptual designs to detailed technical solutions. Most of the content is written at an engineering level providing a lot of details, formulas, charts, and diagrams.

The goal of writing this book is to provide in one place a comprehensive overview of state-of-the-art knowledge on ZEBs to equip architects, engineers, policymakers, and students with the necessary insights needed for their design and construction. This work is a result of many years of teaching and engagement in energy management and sustainability. It reflects synthesis of published research, case studies and personal effort to present recent development in the process of transition to ZEB standards. The book is intended to serve as a resource for students, researchers and professionals dealing with sustainability in architecture and relevant engineering.

The preparation of this book has been greatly enriched by discussions with colleagues and friends. I express my sincere appreciation and gratitude to those who contributed with their suggestions and technical advice. I would also like to thank my institution, the International University of Sarajevo, for providing an academic environment that encourages research. A special and heartfelt gratitude goes to my family for their patience, encouragement, and unwavering support during this journey in challenging moments.

I hope the book will inspire practical solutions and serve as a valuable resource in understanding, designing and implementing of zero-energy buildings.

Sarajevo, Bosnia and Herzegovina

Benjamin Duraković, Ph.D.
bdurakovic@ius.edu.ba

Competing Interests The author has no competing interests to declare that are relevant to the content of this manuscript.

Contents

Chapter 1
Energy Challenges and Building Sector

Abstract This chapter introduces the global energy challenges including sector-specific consumption patterns focusing on building sector and their role in achieving sustainability goals. It discusses the depletion of nonrenewable resources and the environmental consequences associated with them as well as the importance of renewable energy in meeting the goals. Particular attention is given to the building sector, and role of zero energy building concept in fighting energy challenges and achieving energy efficiency. Through an analysis of technological, economic, and regulatory barriers, it was found that achieving net-zero energy in buildings requires coordinated efforts in technology development, policy adoption, and stakeholder education.

Keywords Energy demand · Fossil fuels · Renewable energy · Energy transition · Energy security · Climate change · Energy efficiency · Sustainable development

1.1 Global Energy Challenges

Rapid economic and population growth over the past decades has led to a significant increase in global energy consumption causing substantial increase in greenhouse gas (GHG) emissions. Fossil fuels are considered as primary source of GHG accounting for about 65% of total GHG emissions. A global focus now is on developing and deploying innovative technologies to mitigate these emissions and climate changes.

Electricity generation still relies on fossil fuel combustion and nuclear power as primary energy sources. Both sources have finite reserves and produce waste or pollutants that are accumulated over time. At current consumption rates, global reserves of coal may last roughly 140 years, natural gas about 50 years, and oil only 55 years [1]. Table 1.1 provides a summary of on nonrenewable statistics and associated challenges.

Coal is a major energy source in electricity generation causing substantial environmental GHG emissions. Oil and gas are still in high demand in developing countries but the transition to alternative energy sources is necessity due to its depletion within

B. Duraković, *Zero Energy Buildings*, Green Energy and Technology,
https://doi.org/10.1007/978-3-032-16027-0_1

Table 1.1 Energy sources overview

Energy source	Key points	Statistics	Challenges	References
Coal	• Fastest growing energy source • Key component of energy mix in some countries • Limited reserves	• Fuels 38% of world's electricity • South Africa: 94% of electricity • Poland: 81% of electricity • Reserves deplete in 140 years	• High environmental impact • Limited reserves	[1–4]
Oil	• High demand growth, especially in transportation • Limited reserves	• Reserves deplete in 55 years • Demand growth rate: > 2.0% per year • Critically low in 40 years	• High GHG emissions • Dependency for industrial applications	[1, 2, 5, 6],
Natural gas	• Fastest growing fossil fuel • Significant growth in non-OECD countries	• Reserves deplete in 50 years • Industrial use growth: 1.5% per year • Electric power sector growth: 2.0% per year	• Finite resource • Infrastructure development needed in non-OECD countries	[1, 2, 6],
Nuclear power	• Relies on finite resources • Produces waste products	• Contributes to significant quantities of waste and pollutants	• Nuclear waste management • Safety concerns	[1, 7, 8]
Renewable energy	• Expected to overtake gas and coal by 2040 • Key to reducing fossil fuel dependence	• Renewables projected to become the most used form of primary energy as of 2040	• Initial high investment • Intermittency and storage challenges	[3, 9, 10]
General trends	• Urgent need for alternative energy sources • Importance of energy-saving programs and innovations	• World oil consumption critical in 40 years • Need to preserve fossil fuels for industrial applications	• Transitioning to sustainable energy sources • Reducing building energy demand	[11–14]

the next few decades. In non-OECD countries, gas is still the fastest-growing fossil fuel even though the challenge of finite reserves and infrastructure needs [15]. Nuclear power is in use in various capacities despite its issues with waste management and limited availability of fuel sources.

The expected increase in global energy demand is about 25% by 2040, leading to potential energy shortages and geopolitical tensions. This requires global actions to diversify energy sources and improve infrastructure [16]. Renewable energy sources are also expected to surpass fossil fuels by 2040, despite their discontinuity and high initial cost [17]. Therefore, the need for innovative solutions that will reduce the costs of renewable systems and reduce reliance on fossil fuels are of a high priority [11, 15]. Therefore, energy challenges are multifaceted, but the most critical one is ensuring an uninterrupted energy supply. Table 1.2 shows a broad spectrum of global challenges and their impacts.

As outlined in the Paris Agreement, the transition from fossil fuels to renewable energy is essential to limit global warming to 1.5 °C above pre-industrial levels [18]. Significant obstacles are in this transition process from technological advancements and infrastructural upgrades to financial investments. It is required that investments in renewable energy must be tripled or approximately $5 trillion annually by 2030 or at least of 16.4% annual growth rate [20]. Therefore, a lot of effort is required in developing innovative cost-effective renewable energy solutions along with relevant policies and international cooperation.

Table 1.2 Overview of global energy challenges

Challenge	Description	Impact	References
Energy security	Ensuring a reliable and uninterrupted supply of energy to meet current and future demand	Risk of energy shortages, geopolitical tensions, economic instability	[16]
Climate change	Reducing greenhouse gas emissions from energy production and consumption to mitigate global warming	Rising temperatures, extreme weather events, loss of biodiversity	[18]
Transition to renewable energy	Shifting from fossil fuels to renewable energy sources such as solar, wind, and hydro power	Technological, financial, and infrastructural challenges	[10, 19–21]
Energy efficiency	Improving the efficiency of energy use across various sectors to reduce consumption and emissions	Reduced energy waste, lower costs, decreased environmental impact	[22]
Innovation and technology adoption	Encouraging the development and adoption of new technologies to improve energy production, storage, and consumption	Accelerated transition to sustainable energy, enhanced grid stability, economic opportunities	[23]

Table 1.3 BAS versus smart metering energy flexibility and demand response

Aspect	Building automation systems	Smart metering	Challenges in ZEB
Primary function	Centralized control and automation of building systems	Real-time measurement and communication of energy consumption	Integrating advanced technologies for optimized performance
Key components	HVAC, appliances, lighting, renewable generation, energy storage and other building systems	Advanced metering devices and communication infrastructure	High initial costs for advanced materials and technologies
Data focus	Operational data for system control and adjustments	Energy consumption data for analysis and billing	Ensuring accurate and comprehensive data collection
Automation	High level of automation for controlling building systems	Limited to data collection and communication	Achieving seamless integration of automation and control systems
User interaction	Centralized interface for building users	Data access for utility providers and end-users	Ensuring user engagement and understanding of ZEB benefits
Energy efficiency	Optimizes system operations to minimize energy use	Provides data to identify energy-saving opportunities	Balancing energy generation and consumption in real-time
Real-time adjustments	Automatic adjustments based on predefined parameters	No direct control, but provides data for informed decision-making	Adapting to dynamic energy demands and supply conditions
Integration	Integrates multiple building systems for cohesive management	Integrated with BAS for enhanced energy management	Integrating diverse renewable energy sources and storage systems
Cost savings	Achieved through optimized operation of building systems	Achieved through detailed energy use analysis and informed actions	High upfront investment and long payback periods

1.2 Sector-Specific Energy Consumption

Final energy consumption can be categorized into four main sectors including buildings, industry, transportation and other consumers. Each section exhibits distinct consumption patterns and environmental impacts. Figure 1.1 shows final energy consumption per these sectors based on data extracted from reference [4].

Table 1.4 ZEB challenges

Category	Challenge	Description	Possible solutions
Economic	High initial costs	The upfront investment for ZEBs can be significantly higher due to the cost of advanced materials, technologies, and renewable energy systems	Financial incentives, subsidies, long-term cost savings, and lifecycle cost analysis to justify the investment
	Funding and incentives	Securing funding and understanding available incentives can be difficult for developers and owners	Providing clear guidelines on funding opportunities, simplifying the application process for incentives, and offering financial planning assistance
Technical	Technological integration	Integrating various technologies (solar panels, energy storage, smart grids) and ensuring their compatibility can be complex	Standardization, thorough planning, and the use of integrated building management systems
	Performance verification	Ensuring that buildings perform as designed and achieve zero energy status can be challenging	Continuous monitoring, commissioning, and performance contracts to ensure buildings meet their energy goals
	Energy supply reliability	Dependence on renewable energy sources can be problematic due to variability and intermittency	Integrating energy storage solutions, grid interaction, and diversified energy sources
	Maintenance and operation	The complexity of systems in ZEBs can make maintenance and operation more demanding	Developing comprehensive maintenance plans, training for facility managers, and incorporating user-friendly building management systems
	Innovation and adaptability	Rapid technological advancements require ZEB designs to be adaptable and future-proof to avoid obsolescence	Designing flexible systems, incorporating modular components, and keeping abreast of emerging technologies and practices
Regulatory	Regulatory barriers	Existing building codes and regulations may not support or recognize the standards needed for ZEBs	Advocating for policy changes, updating codes, and creating new standards specific to ZEBs
Practical	Limited expertise	Lack of skilled professionals with experience in designing and constructing ZEBs can hinder development	Training programs, certifications, and increasing awareness among architects, engineers, and builders

(continued)

Table 1.4 (continued)

Category	Challenge	Description	Possible solutions
	Occupant behavior	The energy performance of ZEBs can be significantly affected by the behavior and energy use patterns of occupants	Education and engagement programs for occupants, smart technologies for monitoring and controlling energy use
	Market acceptance	There may be resistance from stakeholders due to unfamiliarity with ZEBs and perceived risks	Demonstration projects, success stories, and clear communication of benefits to build trust and acceptance
	Geographical limitations	The feasibility of achieving zero energy can vary significantly based on local climate, solar exposure, and other geographical factors	Customizing design solutions for local conditions, utilizing site-specific renewable energy sources, and considering microclimate design strategies

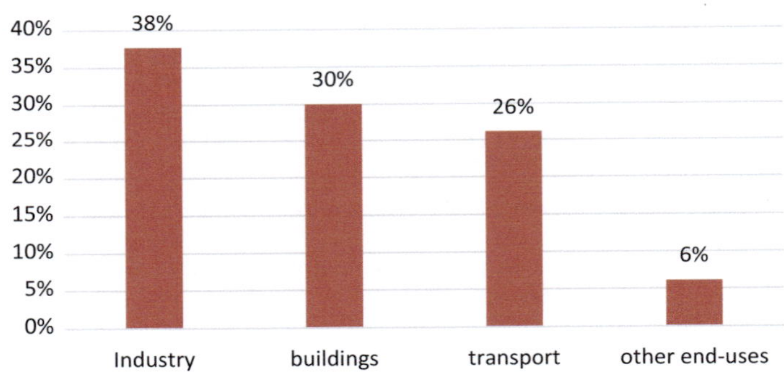

Fig. 1.1 World final energy consumption by sector (Author: based on data in [4])

The industrial sector is the largest consumer of energy, accounting for approximately 38% of global energy use. This high demand is driven by the energy-intensive nature of manufacturing processes, particularly in heavy industries and chemical production. The residential and commercial sectors collectively represent another major consumer with around 30% of the world's energy. Energy use in buildings is driven by heating, cooling, lighting, and appliances. Efforts to reduce building energy demand are directed towards research and development of innovative technologies aimed at lowering fossil fuel consumption. The efforts reflect energy efficiency improvements through enhanced building envelopes, such as better thermal insulation, use of energy-efficient appliances and the integration of smart building technologies.

1.3 Energy Consumption in Buildings and ZEB Concept

Buildings account for approximately 37% of global energy consumption (including embedded energy in material). Therefore, building sector became a primary target for development and application of climate policies worldwide. These policies are aiming to reduce heat loss through the building envelope and enhancing overall building efficiency through the employment efficient HVAC systems, appliances, lighting and financial incentives.

The literature identifies various barriers to energy efficiency investments and proposes different programs and policies to address them. One such approach is the concept of Zero Energy Buildings (ZEBs), which aims to reduce building energy demand through high thermal performance of the building envelope and the integration of on-site renewable energy generation. The ultimate goal of ZEB concept is to achieve a balance between consumed and produced energy. This is achieved by implementing various energy carriers such as building thermal mass, heat storage tanks (heat bank), batteries, photovoltaic, combined cooling, heating and power (CCHP) [24]. The illustration of a conceptual ZEB building is presented in Fig. 1.2.

Policies supporting ZEB concept include building codes and standards, energy performance certification and labeling, financial incentives (grants and tax credits) to mitigate the high initial costs of ZEB technologies, and educational initiatives aimed at stake holders. Improving the energy efficiency of existing buildings is a prerequisite for achieving ZEB status. This can be accomplished through various strategies that reduce building energy demand, including optimized design, enhanced insulation, and intelligent infrastructure. On the other hand, building energy demand response (DR) allows end-users to change their energy consumption behavior as a response to power grid status [25]. This concept is enabled by technologies such as Building Automation (BAS) system and smart metering (SM). BAS refers to the automation and control infrastructure that allows occupants to access, monitor, and

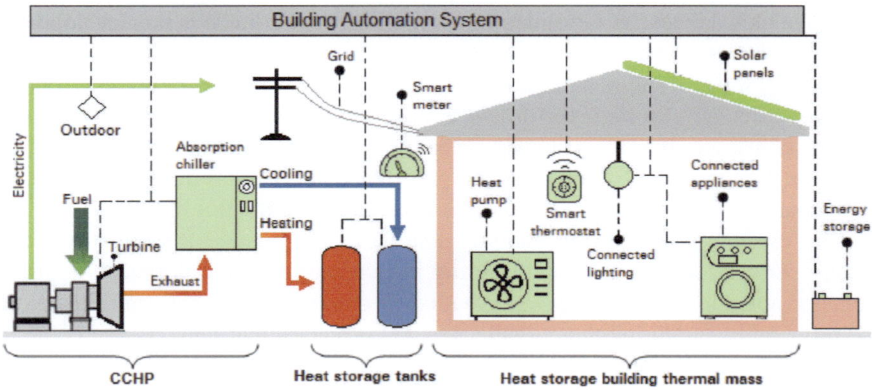

Fig. 1.2 General scheme of ZEB concept. *Source* author

regulate multiple building systems, including HVAC, appliances, lighting, combined cooling, heating and power (CCHP), renewable energy generation, and energy storage. Smart meters, in turn, capture real-time energy consumption data, enabling accurate monitoring, detailed analysis, and efficient energy management (Table 1.3).

BAS is a centralized network that integrates multiple building systems into a unified interface. This way, BAS makes certain adjustments based on real-time data and predefined parameters. It enables automated control and optimization of building systems such as HVAC, lighting, and other key energy-consuming components, while SM provides real-time data on energy usage. Energy Flexibility and Demand Response (EFDR) strategies are integral parts of BAS that enhance the building ability to optimize energy consumption and participate in demand response programs and grid management.

Achieving ZEB status for buildings is a complex and challenging task with a range of specific challenges that need to be addressed. These include technical, regulatory, economic, and practical barriers that must be overcome to enable successful implementation [26, 27]. Technical challenges related to ZEB are associated with achieving buildings energy efficiency without compromising their functionality and user comfort, which require advanced engineering and precise planning [28]. The use of passive strategies such as insulation [29], building shape and height [30], ventilation and thermal mass [31] plays a significant role in this process. Economic challenges are primarily linked to the high initial costs of advanced materials and technologies, as well as the expenses associated with workforce training and the adaptation of construction practices (Table 1.4).

Regulatory frameworks are associated with the lack of regulations, guidelines, and standards, which makes it difficult to implement these technologies. Practical challenges are associated with lack of education and training for all stakeholders within the industry. Therefore, achieving net zero energy consumption in buildings requires a comprehensive approach that includes technological innovation, regulatory reforms, economic incentives, and education of stakeholders [27].

Reducing a building dependence on primary energy has long been the focus of extensive global research. Considerable progress has been made in the development and deployment of energy-saving technologies enabling the integration of renewable energy sources within building environment [32]. Therefore, ZEBs represent an innovative concept for high-performance buildings that integrate renewable energy into the construction process along with energy-efficient building technologies.

Given the complexity of this issue, the objective of this book is to address key technical challenges by providing advances in enabling technologies for net zero buildings by exploring advances in building materials [33], renewable energy systems, energy storage technologies, and offering insight into their development, application, and potential to revolutionize the built environment. This chapter aims to define the problem by exploring these challenges and obstacles that must be overcome to widespread adoption of ZEBs.

1.4 Conclusion

The chapter discussed global energy challenges and the transition toward renewable sources. The building sector emerges as a major consumer of energy and as well a key opportunity for reduction in global demand. Therefore, ZEBs approach embodies a comprehensive solution that integrates high-performance building design, renewable energy generation, and intelligent control systems. The challenges that are identified for a widespread adoption are technical limitations, high initial costs, inadequate regulatory support, and a lack of professional expertise. Addressing these obstacles requires long-term planning, innovative technologies and financing models, as well as enhanced public awareness.

References

1. Years of fossil fuel reserves left. Accessed 17 Jul 2024 [Online]. Available: https://ourworldindata.org/grapher/years-of-fossil-fuel-reserves-left?time=1980..2020
2. Keshavarzian M, Kamali Anaraki S, Zamani M, Erfanifard A (2012) Projections of oil demand in road transportation sector on the basis of vehicle ownership projections, worldwide: 1972–2020. Econ Model 29(5):1979–1985. https://doi.org/10.1016/J.ECONMOD.2012.06.009
3. BP Full report—BP Statistical Review of World Energy 2019
4. World Energy Outlook 2023 (2023). https://doi.org/10.1787/827374A6-EN
5. Sorrell S, Speirs J, Bentley R, Brandt A, Miller R (2010) Global oil depletion: a review of the evidence. Energy Policy 38(9):5290–5295. https://doi.org/10.1016/J.ENPOL.2010.04.046
6. Sorrell S, Speirs J, Bentley R, Miller R, Thompson E (2012) Shaping the global oil peak: a review of the evidence on field sizes, reserve growth, decline rates and depletion rates. Energy 37(1):709–724. https://doi.org/10.1016/J.ENERGY.2011.10.010
7. Petrov VA, Ojovan MI, Yudintsev SV (2023) Material aspect of sustainable nuclear waste management. Sustainability 2023 15(15):11934. https://doi.org/10.3390/SU151511934
8. Mostečšak A, Bedeković G (2018) Metal waste management and recycling methods in the nuclear power plant decommissioning and dismantling process. Rudarsko-geološko-naftni zbornik 33(1):25–33. https://doi.org/10.17794/RGN.2018.1.4
9. Egli F (2020) Renewable energy investment risk: an investigation of changes over time and the underlying drivers. Energy Policy 140:111428. https://doi.org/10.1016/J.ENPOL.2020.111428
10. Alam MS, Al-Ismail FS, Abido MA, Salem A (2020) High-level penetration of renewable energy with grid: challenges and opportunities. IEEE Access 8:190277–190299. https://doi.org/10.1109/ACCESS.2020.3031481
11. Qudrat-Ullah H (2024) The path forward for transition to sustainable energy 203–218. https://doi.org/10.1007/978-3-031-59733-6_9
12. Kabeyi MJB, Olanrewaju OA (2022) Sustainable energy transition for renewable and low carbon grid electricity generation and supply. Front Energy Res 9:743114. https://doi.org/10.3389/FENRG.2021.743114/XML/NLM
13. Ai H, Tan X, Mangla SK, Emrouznejad A, Liu F, Song M (2025) Renewable energy transition and sustainable development: evidence from China. Energy Econ 143:108232. https://doi.org/10.1016/J.ENECO.2025.108232
14. Strielkowski W, Gorina L, Korneeva E, Kovaleva O (2023) Energy-saving technologies and energy efficiency in the post-pandemic world. Accessed 14 May 2025 [Online]. Available: https://arxiv.org/pdf/2312.11711
15. Pablo-Romero P et al (2022) A global assessment: can renewable energy replace fossil fuels by 2050? Sustainability 2022 14(8):4792. https://doi.org/10.3390/SU14084792

16. Energy security in energy transitions—World Energy Outlook 2022—Analysis—IEA. Accessed 18 Jul 2024 [Online]. Available: https://www.iea.org/reports/world-energy-outlook-2022/energy-security-in-energy-transitions
17. Raihan A, Tuspekova A (2023) The role of renewable energy and technological innovations toward achieving Iceland's goal of carbon neutrality by 2040. J Technol Innov Energy 2(1):22–37. https://doi.org/10.56556/JTIE.V2I1.421
18. New Analysis of National Climate Plans: Insufficient Progress Made, COP28 Must Set Stage for Immediate Action | UNFCCC. Accessed 18 Jul 2024 [Online]. Available: https://unfccc.int/news/new-analysis-of-national-climate-plans-insufficient-progress-made-cop28-must-set-stage-for-immediate
19. IRENA—International Renewable Energy Agency. Accessed 18 Jul 2024 [Online]. Available: https://www.irena.org/
20. Tripling Renewables by 2030 Requires a Minimum of 16.4% Annual Growth Rate. Accessed 18 Jul 2024 [Online]. Available: https://www.irena.org/News/pressreleases/2024/Jul/Tripling-Renewables-by-2030-Requires-a-Minimum-of-16-point-4-pc-Annual-Growth-Rate
21. Heti K, Silvanskaya G, Khambir V, Harkava V, Ivanchenko K (2025) The analysis of engineering adaptation to global challenges and strategic planning for the future. Sustain Eng Innov 7(1):137–146. https://doi.org/10.37868/SEI.V7I1.ID453
22. Energy Efficiency 2023—Analysis—IEA. Accessed 18 Jul 2024 [Online]. Available: https://www.iea.org/reports/energy-efficiency-2023
23. MIT Energy Initiative. Accessed 18 Jul 2024 [Online]. Available: https://energy.mit.edu/
24. Beaudin M, Zareipour H, Schellenberglabe A, Rosehart W (2010) Energy storage for mitigating the variability of renewable electricity sources: an updated review. Energy Sustain Dev 14(4):302–314. https://doi.org/10.1016/J.ESD.2010.09.007
25. O'connell N, Pinson P, Madsen H, O'malley M (2014) Benefits and challenges of electrical demand response: a critical review. https://doi.org/10.1016/j.rser.2014.07.098
26. Jaysawal RK, Chakraborty S, Elangovan D, Padmanaban S (2022) Concept of net zero energy buildings (NZEB)—a literature review. Clean Eng Technol 11:100582. https://doi.org/10.1016/J.CLET.2022.100582
27. Attia S et al (2017) Overview and future challenges of nearly zero energy buildings (nZEB) design in Southern Europe. Energy Build 155:439–458. https://doi.org/10.1016/J.ENBUILD.2017.09.043
28. Ascanio-Villabona J, Orejarena-Osorio N, Castillo-Leon NY, Tarazona-Romero BE, Jaimes-Quintero KT (2024) Thermal comfort evaluation using different glazing technologies in an energy model of rural housing in a cold climate. Herit Sustain Dev 6(2):735–748. https://doi.org/10.37868/HSD.V6I2.922
29. Yildiz G, Duraković B, Abd Almisreb A (2021) Performances study of natural and conventional building insulation materials. Int J Adv Sci Eng Inf Technol 11(4):1395–1404. https://doi.org/10.18517/IJASEIT.11.4.11139
30. Hameed AI, Kakoo HS, Musa VA, Abdulla SI, Ebrahimpou A (2024) A multi-sectional analysis of building height, layout, and urban density on seasonal energy consumption: a case study. Sustain Eng Innov 6(2):167–176. https://doi.org/10.37868/SEI.V6I2.ID308
31. Duraković B (2020) Passive solar heating/cooling strategies. In: PCM-based building envelope systems: innovative energy solutions for passive design, Springer Nature, pp 39–62. https://doi.org/10.1007/978-3-030-38335-0_3
32. Jacobson MZ (2009) Review of solutions to global warming, air pollution, and energy security. https://doi.org/10.1039/b809990c
33. Findik F (2022) Green concrete for structural buildings. Herit Sustain Dev 4(1):67–76. https://doi.org/10.37868/HSD.V4I1.84

Chapter 2
Fundamentals, Core Principles, and Trends of Zero Energy Buildings

Abstract This chapter introduces the foundational concepts of Zero Energy Buildings (ZEBs), focusing on their definition, energy balance principles, and various international standards and certifications. It highlights the limits of energy use in ZEBs and the importance of ZEBs in mitigating urban energy consumption and greenhouse gas emissions, with an emphasis on energy-efficient technologies and renewable energy integration. The discussion covers regulatory frameworks and some trends and challenges associated with achieving net zero energy status in buildings worldwide. To drive the adoption of ZEBs, global cooperation is essential to implement effective policies, harmonize standards, provide education, and foster innovation.

Keywords Zero energy buildings · Definition and concepts · Energy balance · ZEB standards harmonization · Global cooperation · Renewable energy integration

Abbreviations and Acronyms

AI	Artificial Intelligence
ASHRAE	American Society of Heating, Refrigerating and Air-Conditioning Engineers
EEA	European Environment Agency
EPB	Energy-plus buildings
EPBD	Energy Performance of Building Directive
G	Energy generation
GHG	Greenhouse gasses
HVAC	Heating, Ventilation, and Air Conditioning
IoT	Internet of Things
L	Load
ML	Machine Learning
nZEB	Near Zero-Energy Building
NZEB	Net Zero-Energy Building
ZEB	Zero-Energy Building

B. Duraković, *Zero Energy Buildings*, Green Energy and Technology,
https://doi.org/10.1007/978-3-032-16027-0_2

2.1 Introduction

The global population has reached a total of eight billion people, which is a significant milestone in terms of development. The increase in world population requires an increase in housing and energy. As of 2020, the European Environment Agency (EEA) reported that around 74% of the population in the European Union (EU) lives in urban areas [1]. Extensive urbanization has caused an increase in energy consumption for buildings, transport and industry, which has led to significant greenhouse gas emissions [2]. According to estimates, 70% of energy-related greenhouse gas emissions are attributed to urban areas in Europe [3]. The construction industry is the most significant contributor to today's energy consumption. Its energy demand is approximately 40% of global energy use, while it accounts for 36% of the European Union's CO_2 emissions [4, 5]. Therefore, buildings account for a significant part of global energy consumption and carbon dioxide emissions, making them an imperative area for research and application of technologies that can transform them into zero-energy buildings (ZEB), also known as net zero-energy buildings (NZEB).

There are various definitions of ZEBs, but the simplest definition could be buildings with exceptionally high energy performance, where nearly zero or a very small amount of the energy required is significantly supplied by renewable sources [6] According to ASHRAE, ZEBs are defined as buildings that consume no more energy on an annual basis than is generated on-site from renewable energy systems [7]. Also referred to as Net Zero Energy Buildings (NZEBs), these structures are designed with the primary goal of increasing energy efficiency by minimizing overall energy demand and maximizing on-site renewable energy generation. Although such buildings may occasionally draw energy from the grid, their net annual energy consumption is significantly reduced. NZEBs typically achieve this balance by implementing passive design strategies, high-efficiency performance building envelopes, and advanced energy-efficient technologies.

2.1.1 *Energy Balance and ZEB Categories*

To define ZEB, the concept of energy balance concept is key in this process. Import/export energy balance and the load/generation energy balance are proposed by Sartori et al. [8]. The import/export energy balance assesses the ability of the building to meet the own load by on-site generation, while the load/generation energy balance assesses the ability of the building to meet its own energy demand and energy generated from on-site renewable sources at the end of the billing period, which can be useful during the design phase. Figure 2.1 presents the relevant terminology within the building system boundary for the NZEB balance concept. The strategy is based on Sartori et al. [8].

A *weighting system* is a method used to assign relative importance to different components or factors within energy generation and consumption. In addition, the

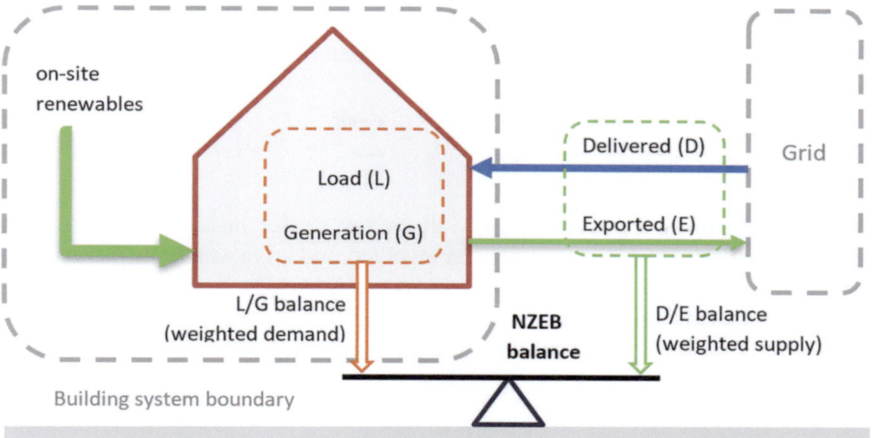

Fig. 2.1 NZEB balance in terms of load/generation or delivered/exported energy exchanges

weighting factors may reflect engineering considerations as well as political preferences. It prioritizes and balances different sources and helps to assess the energy performance of a building. On the *energy consumption* side, weights can be assigned to energy-consuming activities including heating, cooling, lighting, and appliances. These weights are assigned based on the energy consumption and efficiency of a particular energy-consuming activity in the total energy demand. On the *energy generation* side, weights can be assigned to different energy sources including solar panels, wind turbines, and geothermal systems. These weights are assigned based on the contribution of a particular energy carrier to the total energy production. In this regard, the terminology used in Fig. 2.1 has the following meanings:

- *Weighted demand* is the total energy consumption of a building, adjusted by weights assigned to different energy-consuming activities or systems within the building. This helps to prioritize the various components of energy demand based on their significance.
- *Weighted supply* is the total energy generated by the building from various sources, adjusted for the weights assigned to different energy generation carriers. This represents the contribution of each energy source to the total energy supply.
- *Load* is the building energy demand per energy carrier in (kWh/y) or (kWh/m^{2}y). Due to on-site self-generation and energy consumption, the energy produced on-site from renewable sources may not match the exported energy.
- *Generation* means on-site energy generation from renewable sources per energy carrier in (kWh/y) or (kWh/m^{2}y). Due to own energy consumption, on-site-generated energy from renewables may not match the exported energy.
- *Delivered energy* is the energy flow from the grid to the building per energy carrier in (kWh/y) or (kWh/m^{2}y).
- *Exported* is the energy flow from the building to the grid per energy carrier in (kWh/y) or (kWh/m^{2}y).

The globally accepted concept of energy balance for NZEBs is the balance between weighted demand and weighted supply described in Eq. (2.1).

$$E_{bal} = \sum_{i=1}^{n} E_{s,i} \cdot w_{s,i} - \sum_{j=1}^{m} E_{L,j} \cdot w_{L,j} \tag{2.1}$$

where, E_s is energy supplied by any i-th energy carrier involved, E_L is energy demanded (load) by any j-th energy carrier involved, and w is a weighting factor incorporated in the balance to account for different energy types. E_{bal} is the energy quantity, and its unit depends on the chosen weighting system (it is usually in kWh). Energy quantities for supplied and exported energy can be calculated using the following balance equation:

$$\sum_{i=1}^{n} E_{e,i} \cdot w_{e,i} - \sum_{j=1}^{m} E_{d,j} \cdot w_{d,j} = E - D \geq 0 \tag{2.2}$$

where, E_e and E_d are exported and delivered energy, respectively; w is the weighting factor for each and i-th and$j$-th energy carrier. E is the weighted exported energy, and D is the weighted delivered energy. Since in most cases only generated energy and building energy demand (load) are available, therefore *load/generation balance* can be calculated as follows:

$$\sum_{i=1}^{n} E_{g,i} \cdot w_{g,i} - \sum_{j=1}^{m} E_{L,j} \cdot w_{L,j} = G - L \geq 0 \tag{2.3}$$

where, E_g and E_L stand for energy generation and load, respectively, w is the weighting factor for each i-th and$j$-th energy carrier. G is the weighted generation, and L is the weighted load. The *import/export balance* between weighted exported and delivered energy, and *load/generation balance* between weighted generation and load are shown graphically in Fig. 2.2.

Equation (2.4) represents the conceptual definition of ZEB:

$$\begin{aligned} &\textit{nearly ZEB if } E_{bal} \lesssim 0 \\ &\textit{ZEB if } E_{bal} = 0 \\ &\textit{plus ZEB if } E_{bal} > 0 \end{aligned} \tag{2.4}$$

During the planning and design phase of ZEB, it was necessary to use renewable energy, implement passive design approaches, and select energy-efficient active systems. Reducing energy demand in a building without reducing occupant comfort is a prerequisite for ZEB and is primarily achieved through energy efficiency and energy savings by insulating the building envelope, protecting it from the sun and using equipment and systems with improved efficiency.

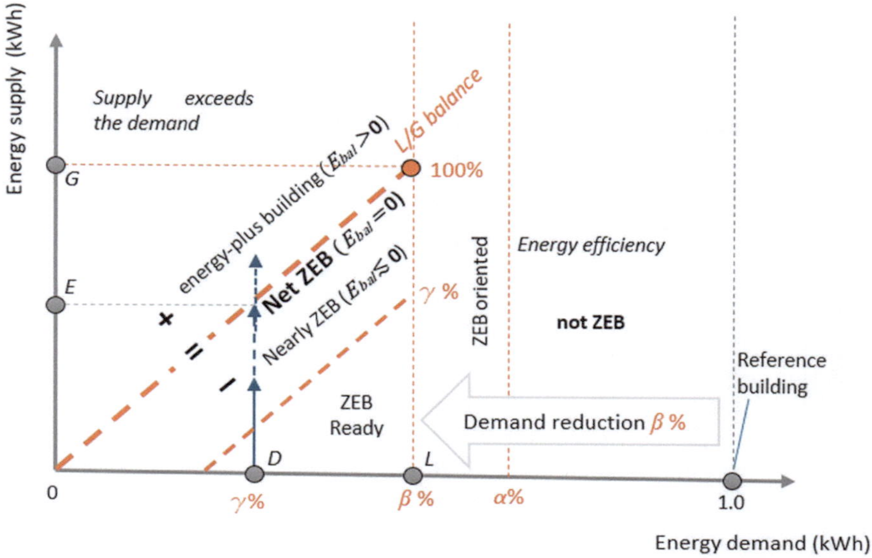

Fig. 2.2 ZEB balance concept

Climatic conditions, building topology, and national regulations play an important role in achieving ZEB. National regulations around the world recognize different ZEB categories. Analyzing these regulations and the literature, several building categories are identified. Therefore, ZEB-certified buildings can be ranked from highest to lowest as Energy-plus buildings (EPB), ZEB, Nearly ZEB, ZEB Ready, and ZEB Oriented (Fig. 2.3). ZEB Oriented are ranked only in Japan by The Building-Housing Energy-efficiency Labeling System (BELS), which is a certification system that evaluates and labels the energy efficiency of buildings [9]. Summary of the definition for each ZEB category is provided in Table 2.1. The table provides a clear and concise overview of each category in the Zero Energy Building spectrum, including the newly added EPB, with both qualitative and quantitative definitions.

In Table 2.1, the percentages are presented using variables (α, β, γ) to denote their common use in different scenarios, as well as typical quantitative values defined in national regulations. This provides a comprehensive quantitative understanding of each rank. A graphical visualization of the ZEB classification is shown in a bar chart diagram that highlights the different levels of energy performance (see Fig. 2.3).

The above bar chart visualizes the incremental improvements in energy performance standards from ZEB Oriented to Energy-plus building.

- A ZEB-oriented building is a building that meets the α% reduction in energy demand compared to a reference building. This category is recognized in some national regulations such as those of Japan, and the building is certified as a ZEB-oriented building.

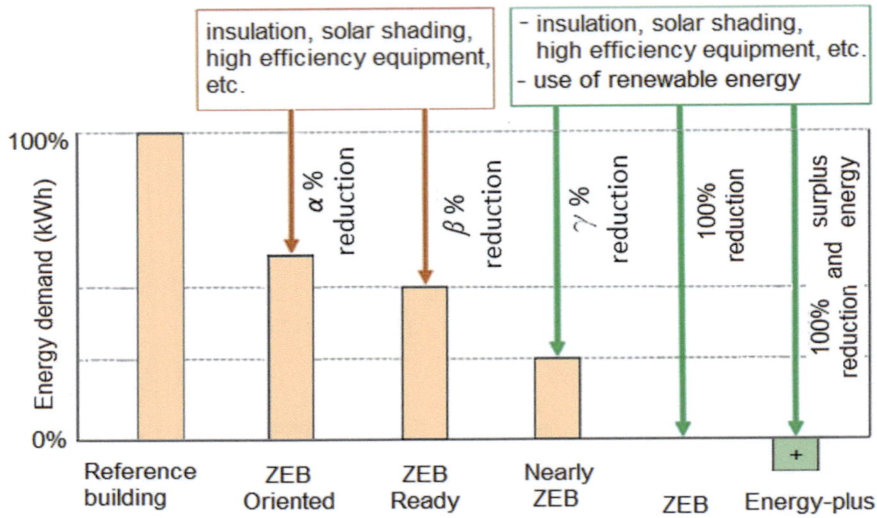

Fig. 2.3 Energy balance and ZEB categories

Table 2.1 Summary of ZEB ranking definitions

Rank	Qualitative definition	Quantitative definition
ZEB oriented*	Aim to be ZEB ready, with measures for achieving energy savings using thermal insulation of building envelope, energy-efficient equipment, etc.	Buildings with energy demand reduction of at least α% (usually 30% in hotels and 40% in office buildings or more) with the use of promising technology
ZEB ready	Use advanced architectural designs aiming to become ZEBs, with a highly energy efficient thermal insulation of the building envelope, energy-efficient equipment, etc.	Buildings with primary energy demand reduction of at least by β% (usually 50% or more)
Nearly ZEB	Have reduced annual primary energy use to nearly zero, by using own energy generation from renewables	Buildings with annual primary energy demand reduction of at least of γ% (usually 75% or higher compared to standard buildings) and have at least 50% or higher energy savings and energy generation from renewable sources
ZEB	Achieved annual primary energy use to net zero	Buildings with annual primary energy demand reduction of 100% or more and have at least 50% or higher energy savings and energy generation from renewable sources
EPB	Produce surplus energy, contributing to energy supply beyond their own needs	Buildings that generate more energy than they consume, typically achieving more than 100% net energy saving, effectively exporting surplus energy

* In Japan only

- A ZEB-ready building is a building that achieves at least β% reduction in primary energy demand compared to a reference building.
- A nearly ZEB building is a building that achieves at least γ% of primary energy demand reduction relative to a reference building.
- A (net) ZEB building is a building that achieves 100% of primary energy demand reduction relative to a reference building.
- A ZEB Plus known as energy-plus building (or plus energy building) is an advanced concept in the spectrum of energy-efficient buildings. It goes beyond the standard ZEB criteria by generating on a yearly basis more energy from renewable sources than it imports from the grid.

2.1.2 Boundaries of Energy Use

The boundary of building system defines the scope of energy generation and consumption activities and includes physical boundary and balance boundaries. This boundary is essential to ensure that all relevant energy flows in and out of the system are included in the balance. A well-defined boundary helps to understand the energy dynamics of a building and contributes to better decision-making.

Physical boundary: refers to the tangible limits of a building infrastructure and building systems, and includes physical components and location where energy is generated, consumed, or stored. That includes:

- Building envelopes—all elements of a building that separate the interior from the exterior (walls, roofs, windows, and doors…), which are used to determine building thermal performance.
- On-site generation: renewable energy systems (solar panels, wind turbines, and geothermal systems) and non-renewable systems (backup generators).
- On-site energy consumption: include HVAC systems, lighting systems, household appliances, office equipment, and water heating systems.
- Energy storage systems: this boundary includes batteries for electrical energy and thermal storage as physical and latent heat for heating or cooling.

Balance boundary refers to a perimeter within which all energy flows are measured and calculated including supply, demand, generation, export, and storage. This boundary includes physical elements mentioned above and virtual elements that influence energy balance. Virtual elements that influence energy balance include:

- Energy management and monitoring systems that monitor, manage, and optimize energy flows within the boundary.
- Building Automation Systems that integrate various building systems (e.g., HVAC, lighting, security) to work together efficiently, automate routines based on occupancy, time of day, and remote management that monitor and control the system.
- Smart Metering and Sensors collect data and provide Real-Time Feedback on energy use enabling immediate adjustments.

Table 2.2 Comparison of physical boundary versus balance boundary

Aspect	Physical boundary	Balance boundary
Definition	Tangible, spatial limits of the building's infrastructure and systems	Defined perimeter for measuring and accounting all energy flows
Components included	Building envelope, on-site energy generation and consumption systems, energy storage, grid connection points	Energy generation, consumption, imports, exports, and storage; includes both physical and virtual elements
Purpose	Defines the physical limits and infrastructure of the building	Ensures accurate energy accounting, performance evaluation, and optimization
Examples	Walls, roofs, windows, doors, HVAC systems, lighting, appliances	On-site renewable energy generation, energy imports/exports, energy storage, energy management systems
Relevance to energy assessment	Focuses on physical components and spaces affecting energy flows	Focuses on the complete picture of energy dynamics within the defined boundary for performance assessment

- Energy Analytics Platforms provides data analysis and reports on energy savings opportunities in real-time.
- Demand Response Systems includes load management and incentives, which adjust energy usage during peak periods to reduce demand and lower costs.
- Artificial Intelligence (AI) and Machine Learning (ML) are algorithms that from historical data predict energy demand and optimize system operations and continuously improve efficiency based on AI-driven insights.

Understanding the differences between the physical boundary and the balance boundary is essential for managing, assessing, and optimizing the building energy performance. A summary of the boundaries is presented in Table 2.2.

Both boundaries are essential for comprehensive energy management. A physical boundary helps understand the physical spaces and components while the balance boundary ensures all energy dynamics are included in order to facilitate and optimize energy use.

2.2 Trends and Forecasts

According to forecasts, the ZEB market is expected to reach a substantial value by 2031, reflecting a strong compound annual growth rate over the period. It is expected to grow significantly in the coming years. The growth is reflecting the ongoing global shift towards energy-efficient and sustainable building practices. The market value of ZEBs in 2023 was USD 81.9 Billion and it is projected to reach USD 277.9 billion, growing at a compound annual growth rate (CAGR) of 16.5% from 2024 to 2031 [10]. Per another reports the market value in 2022 was USD 27.59 billion with CAGR of 23.2% [11], reaching USD 180.338 billion by 2031.

Driving factors of the growth are increasing government incentives and regulations that promote energy efficiency, growing environmental awareness, and increasing cost of energy.

Financial incentives for adoption of ZEB technologies are promoted across various regions including subsidies, tax credits, and grants to support energy-efficient and sustainable buildings [12]. Technological advancements in renewable energy systems, such as solar panels and energy storage solutions, are crucial in the expansion of the market. Currently the largest market share holds the United States and Canada due to its stringent energy standards and supportive policies. Europe also holds a considerable share, which is driven by the Energy Performance of Buildings Directive (EPBD) and the near ZEB targets [12]. Due to a rapid urbanization and industrialization in countries like China and India, the Asia–Pacific region is expected to grow at the fastest rate [10]. Market overview of trends and challenges from technological, economic, regulatory and environmental aspects is presented in Table 2.3.

ZEBs are designed with the objective of eliminating all greenhouse gas emissions associated with building operations and contributing to climate change mitigation efforts. This objective includes direct emissions from building energy use such as fossil fuels for heating, cooling, and indirect emissions from sources such as transportation or embodied carbon in materials. Therefore, NZEBs keep a balance between energy consumption and on-site renewable energy generation, which results in net zero energy consumption at end of the year. Achieving NZEB status for buildings is a complex and challenging task with several specific issues that need to be addressed. These issues include technical, regulatory, economic and practical challenges [5, 16].

Technical challenges related to NZEB are associated with achieving energy efficiency of buildings without compromising their functionality and user comfort, which requires advanced engineering and precise planning. The use of passive strategies such as efficient glazing technologies [17], insulation [18], ventilation and building shape [19] and thermal mass [20] plays a significant role in this process. Economic challenges are associated with high initial costs of these materials and technologies as well as costs for training the workforce and adapting construction practices. Regulatory frameworks are associated with the lack of regulations, guidelines, and standards, which makes it difficult to implement these technologies. Practical challenges are associated with lack of education and training for all stakeholders within the industry. Therefore, achieving net zero energy consumption in buildings requires a comprehensive approach that includes technological innovation, economic incentives, regulatory reforms and education of industry stakeholders [16].

Table 2.3 ZEB trends and challenges

Category	Market drivers	Market trends	Market challenges
Technological	Technological advancements: innovations in building materials, renewable energy systems, and smart technologies make ZEBs more feasible and cost-effective [12]	Integration of renewable energy: greater adoption of solar PV, wind turbines, and geothermal systems [10, 12] Smart building technologies: increased use of IoT and automation for real-time energy management [12] Retrofitting existing buildings: upgrading existing buildings with energy-efficient technologies [10]	Technical complexity: specialized knowledge and skills required for designing and constructing ZEBs [12, 13] Performance verification: ensuring ZEBs achieve energy performance goals requires consistent monitoring [10, 13]
Economic	Rising energy costs: increasing energy prices motivate building owners to invest in ZEBs to reduce long-term operational expenses [10]	Decreasing costs of renewable technologies: the reduction in the cost of solar panels and other renewable technologies makes ZEBs more financially viable [10]	High initial costs: significant upfront investment required for ZEB technologies and construction practices [10]
Environmental	Environmental awareness: growing consciousness about climate change and the need to reduce carbon emissions boosts demand for sustainable buildings [14]	Holistic sustainability: emphasis on overall sustainability, including water conservation and waste reduction [12]	Market Awareness: lack of widespread understanding of ZEB benefits among potential buyers and investors [12]
Regulatory	Government regulations and incentives: stricter building codes and energy efficiency regulations, financial incentives like tax credits and subsidies [10, 12]	Global policy alignment: increasing alignment of international policies on building energy efficiency [10, 15]	Compliance costs: financial burden of adhering to stringent regulations can be high for builders [10]

2.3 Some of the Global Initiatives and Standards

Global emission reduction goals are promoted through the United Nations Framework Convention on Climate Change (UNFCCC) in the Copenhagen Agreement [21] and the Paris Agreement [14]. The 2015 Paris Agreement includes a commitment by 195 countries to limit global temperature rise to no more than 1.5 °C above pre-industrial levels [22].

The Asia–Pacific Economic Cooperation (APEC) members have set ambitious goals: to reduce energy intensity by 45% in the period from 2005 to 2035 and to double the share of renewable energy in their energy mix in the period 2010 to 2030 [23, 24]. It is worth noting that the APEC region represents approximately 60% of global energy demand [25] positioning it to drive the ongoing rapid changes in the global energy system. Several APEC members have achieved and published successful best practices on ZEB including China, Korea, Japan, Russia, Malaysia, Singapore, Canada, Chile, United States and Australia, [26–30].

Setting NZEB standards is a constant issue and appears to vary considerably from country to country [31]. Those standards are not easy to set, since each country has their own rules and regulations. In recent years, numerous nations, organizations, and groups have released a range of initiatives and political stances in favor of the NZEB research. Task 40, "Towards Net Zero Energy Solar Buildings," was approved by the International Energy Agency's (IEA) Solar Heating and Cooling Program (SHC) in 2008 [32]. Creating industry rules, harmonized guidelines, an international definition framework, instruments, creative solutions, tools, and a shared understanding is the task's goal. Installing the global NZEB database is part of the Task's framework. This is the Global NZEB Database Installation, which is the demonstration of projects in this database and offers precise creation, use, and evaluation of experiences. Thanks to the Energy Independence and Security Act of 2007 (EISA 2007), this NZEB initiative aims to achieve net zero energy for all new commercial buildings by 2030 [33, 34].

European Union's Energy Performance of Buildings Directive (EPBD) [35] on nearly zero-energy buildings (nZEB), has undergone revisions over the years to strengthen energy efficiency measures in buildings across EU member states. The specific requirement mandates that all newly constructed buildings must ensure that the amount of energy they produce on-site, typically through renewable energy sources such as solar panels or wind turbines, matches the amount of energy they consume for heating, cooling, ventilation, lighting, and hot water [35]. The deadline for all new buildings to meet the nZEB standard varies slightly depending on the specific version of the directive and the individual member state's transposition of the directive into national law [33].

All new buildings in the EU must be Zero-Emission Buildings (ZEBs) by 2030. Public buildings are required to meet this standard by 2027. Harmonized standards will be set at the EU level, requiring renovations for buildings with the lowest energy performance certificates to improve to at least class F (*see Sect.* 2.4*of this chapter*) by 2030 for non-residential and by 2033 for residential buildings [35, 36]. Similar programs or promotion ideas for the NZEB or home renewable energy consumption have also been proposed by the UK, Canada, and Japan [37–39]. South Korea's certification regulations further exemplify the growing emphasis on zero-energy buildings (ZEBs) in Asia. In China MoHURD and CABEE standards and Australia National Construction Code (NCC) Section J from 2010 reflect a strong commitment to promoting ultra-low energy and nearly zero energy buildings. Table 2.4 presents a list of some existing voluntary standards and guidelines for ZEB.

Table 2.4 presents an overview of efforts and initiatives towards improving energy efficiency and promoting sustainable building practices through regulatory frameworks and standards for selected regions/countries. The table introduces some of the

Table 2.4 ZEB standards and design guidelines

Region/ country	Organization	Year & Title	Reference
Australia	Australian Building Codes Board	2010 National Construction Code (NCC)—Section J	[40]
European Union	European Commission	2010 Energy Performance of Buildings Directive (EPBD)	[35, 36, 41]
USA	ASHRAE	2018 Advanced Energy Design Guide for K-12 School Buildings	[42]
	ASHRAE	2019 Advanced Energy Design Guide for Small to Medium Office Buildings	[43]
	Architecture 2030	2019 The ZERO Code—A Zero-Net-Carbon International Building Code Standard	[44]
Canada	CHBA	2017 Net Zero Home Labelling Program Technical Requirements	[45]
China	MoHURD	2015 Passive Ultra-low Energy Green Building Technical Guidance (residential)	[46]
	MoHURD	2019 Nearly Zero Energy Building Technical Standard GB/T51350-2019	[47]
	CABEE	2019 Nearly Zero Energy Building Evaluation Standard TB/CABEE 003-2019	[48]
Japan	SHASE	2015 Definitions and Evaluation Method of ZEB	[49]
	METI/SII	2017–2019 ZEB Design Guideline	[50]
South Korea	MOLIT	2017 Regulation on Certification of Building Energy Rating and Certification of ZEBs	[51]
Russia	Ministry of Construction, Housing and Utilities	2016 National Standard for Energy Efficiency in Buildings	[52]
Turkey	Ministry of Energy and Natural Resources	2017 National Energy Efficiency Action Plan	[53]
India	Bureau of Energy Efficiency	2017 Energy Conservation Building Code (ECBC)	[54]
Mexico	CONUEE	2013 Normas Oficiales Mexicanas (NOM) for Energy Efficiency in Buildings	[55, 56]
Norway	Norwegian Ministry of Local Government and Modernization	2017 TEK17 Building Regulations	[57]
Switzerland	Swiss Federal Office of Energy	2018 Minergie Standard	[58]

national commitments to energy efficiency by organizations/governmental bodies within Australia, the European Union, the USA, Canada, China, Japan, South Korea, Russia, Turkey, India, Mexico, Norway, and Switzerland. Overall, these regulatory frameworks and standards are tailored to local contexts and trace a global shift towards more energy-efficient and environmentally sustainable buildings.

2.4 Energy Labelling of Buildings

Energy certificates are tools for promoting energy efficiency and sustainability in buildings. These tools provide a standardized measure of energy performance and practical recommendations for improvement. These certificates help in decision making towards energy conservation and reducing the environmental impact of buildings. The energy certificate is mandatory in many countries when selling or renting a building, apartment or office space, and is valid for ten years, after which it must be renewed. Therefore, the certificate provides information to the owner, future building buyer, or user on how much energy the building consumes.

Basically, an energy certificate is a document that presents the energy properties of the building in a reviewed manner, such as heating, carpentry, quality of walls and climate. This document has prescribed specific content and appearance; it is issued by an authorized person. It states how much energy a building consumes, how much energy it wastes unnecessarily, and which energy class it belongs to. In addition, the certificate contains a proposal for the measures a building owner can take to get the building into a higher energy class with an estimate of energy savings and the time frame in which these measures will pay off [41].

There are eight energy classes, with A+ indicating the most energetically favorable class, and G the least energetically favorable class. The energy class shows the annual energy consumption per unit area in (kWh/m^2a). An example of energy performance certificate is shown in Fig. 2.4.

Class “A+ ” denotes buildings that consume less than 15 kWh/m^2, and class “G” denotes buildings that consume more than 250 kWh/m^2. In class A there are low-energy buildings that have very good insulation, multiple insulated walls, triple glazed PVC carpentry and use renewable sources. Class B includes well-insulated buildings, well protected by neighboring buildings. Class C includes well-insulated buildings and newer apartments, while in class D buildings are insulated from all sides, with PVC joinery, well-insulated apartments on the edges of buildings or poorly insulated apartments surrounded by other apartments.

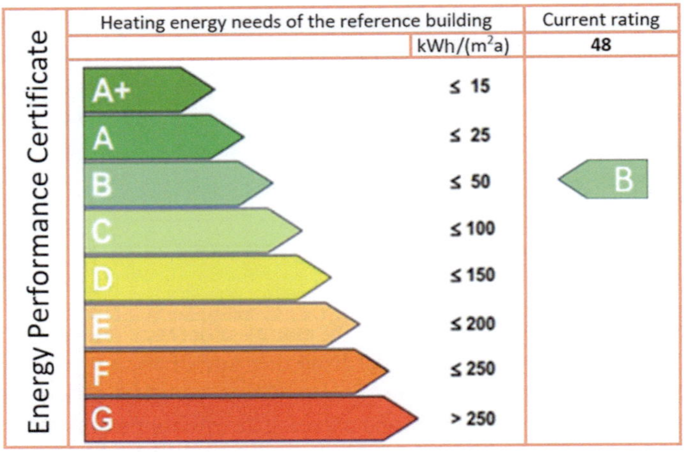

Fig. 2.4 Example of energy performance certificate issued for building owners

2.5 Concluding Remarks

In conclusion, the discussion highlights that ZEBs represent a key path towards sustainable urban development and climate change mitigation. Defined as buildings that produce as much energy as they consume over a year, ZEBs rely on advanced architectural designs, energy-efficient technologies, and on-site renewable energy generation. Despite the different international standards and definitions, the overarching goal remains consistent: reducing carbon footprints and enhancing energy efficiency in the built environment. Challenges such as high initial costs, technical complexity, and regulatory variations require comprehensive approaches involving technological innovation, supportive policies, and stakeholder education. Moving forward, continued global cooperation and alignment on ZEB standards are essential to accelerate adoption and achieve significant environmental benefits on a broader scale.

References

1. Harris S, Mata É, Lucena AFP, Bertoldi P (2023) Climate mitigation from circular and sharing economy in the buildings sector. Resour Conserv Recycl 188:106709. https://doi.org/10.1016/J.RESCONREC.2022.106709
2. Prozuments A, Kim AA, Borodinecs A (2020) Evaluation of the building stock thermal performance under various building code compliance scenarios: the case of Latvia. In: Construction research congress 2020: infrastructure systems and sustainability—selected papers from the construction research congress 2020, pp 532–542. https://doi.org/10.1061/9780784482858.058
3. Kneifel J, Annual whole building energy simulation of the NIST net zero energy residential test facility design. https://doi.org/10.6028/NIST.TN.1767

4. Vanhoudt D, Geysen D, Claessens B, Leemans F, Jespers L, van Bael J (2014) An actively controlled residential heat pump: potential on peak shaving and maximization of self-consumption of renewable energy. Renew Energy 63:531–543. https://doi.org/10.1016/J.RENENE.2013.10.021
5. Jaysawal RK, Chakraborty S, Elangovan D, Padmanaban S (2022) Concept of net zero energy buildings (NZEB)—a literature review. Clean Eng Technol 11:100582. https://doi.org/10.1016/J.CLET.2022.100582
6. D'Agostino D, Mazzarella L (2019) What is a Nearly zero energy building? Overview, implementation and comparison of definitions. J Build Eng 21:200–212. https://doi.org/10.1016/J.JOBE.2018.10.019
7. ASHRAE: ANSI/ASHRAE Standard 228-2023 (2023) Standard method of evaluating zero net energy and zero net carbon building performance
8. Sartori I, Napolitano A, Voss K (2012) Net zero energy buildings: a consistent definition framework. Energy Build 48:220–232. https://doi.org/10.1016/J.ENBUILD.2012.01.032
9. About BELS (building energy efficiency labeling system) | General incorporated association housing performance evaluation and labeling association. https://www.hyoukakyoukai.or.jp/bels/bels.html. Last accessed 05 July 2024
10. In-depth industry outlook: zero energy buildings market size, forecast. https://www.verifiedmarketresearch.com/product/zero-energy-buildings-market/. Last accessed 2024/07/05.
11. Zero-Energy Building (ZEB) Market | Global Industry, 2031. https://www.businessresearchinsights.com/market-reports/zero-energy-building-zeb-market-102943. Last accessed 05 July 2024
12. Zero energy buildings market size, share, growth | Report, 2032. https://www.marketresearchfuture.com/reports/zero-energy-buildings-market-5325. Last accessed 05 July 2024
13. Heti K, Silvanskaya G, Khambir V, Harkava V, Ivanchenko K (2025) The analysis of engineering adaptation to global challenges and strategic planning for the future. Sustain Eng Innov 7:137–146. https://doi.org/10.37868/SEI.V7I1.ID453
14. Unfccc: adoption of the Paris agreement—Paris agreement text English
15. Al-Sukkari TAS, El-Daghar K, Galal KS, Afify A (2024) Towards effective implementation of building energy efficiency codes in Tripoli, Lebanon: key actions for enforcement. Herit Sustain Dev 6:771–790. https://doi.org/10.37868/HSD.V6I2.675
16. Attia S, Eleftheriou P, Xeni F, Morlot R, Ménézo C, Kostopoulos V, Betsi M, Kalaitzoglou I, Pagliano L, Cellura M, Almeida M, Ferreira M, Baracu T, Badescu V, Crutescu R, Hidalgo-Betanzos JM (2017) Overview and future challenges of nearly zero energy buildings (nZEB) design in Southern Europe. Energy Build 155:439–458. https://doi.org/10.1016/J.ENBUILD.2017.09.043
17. Ascanio-Villabona J, Orejarena-Osorio N, Castillo-Leon NY, Tarazona-Romero BE, Jaimes-Quintero KT (2024) Thermal comfort evaluation using different glazing technologies in an energy model of rural housing in a cold climate. Herit Sustain Dev 6:735–748. https://doi.org/10.37868/HSD.V6I2.922
18. Yildiz G, Duraković B, Abd Almisreb A (2021) Performances study of natural and conventional building insulation materials. Int J Adv Sci Eng Inf Technol 11:1395–1404. https://doi.org/10.18517/IJASEIT.11.4.11139
19. Hameed AI, Kakoo HS, Musa VA, Abdulla SI, Ebrahimpou A (2024) A multi-sectional analysis of building height, layout, and urban density on seasonal energy consumption: a case study. Sustain Eng Innov 6:167–176. https://doi.org/10.37868/SEI.V6I2.ID308
20. Duraković B (2020) Passive solar heating/cooling strategies. In: PCM-based building envelope systems: innovative energy solutions for passive design. Springer Nature, pp 39–62. https://doi.org/10.1007/978-3-030-38335-0_3
21. UNFCCC (2009) Copenhagen accord—U.N. framework convention on climate change
22. Moran P, Goggins J, Hajdukiewicz M (2017) Super-insulate or use renewable technology? Life cycle cost, energy and global warming potential analysis of nearly zero energy buildings (NZEB) in a temperate oceanic climate. Energy Build 139:590–607. https://doi.org/10.1016/J.ENBUILD.2017.01.029

23. 2011 APEC Ministerial Meeting | APEC. https://www.apec.org/Meeting-Papers/Annual-Ministerial-Meetings/2011/2011_amm. Last accessed 25 June 2024
24. APEC energy demand and supply outlook 7th edition-volume I APEC energy working group (2019)
25. Taniguchi M, Masuhara N, Burnett K (2017) Water, energy, and food security in the Asia Pacific region. J Hydrol Reg Stud 11:9–19. https://doi.org/10.1016/J.EJRH.2015.11.005
26. Zhang S, Zhang YJ (2019) Erratum: proposal of a new nomenclature for introns in protein-coding genes in fungal mitogenomes. IMA Fungus 10(15). https://doi.org/10.1186/s43008-019-0015-5; IMA Fungus (2020) 11:7. 10.1186/s43008-020-00030-2
27. Feng W, Zhang Q, Ji H, Wang R, Zhou N, Ye Q, Hao B, Li Y, Luo D, Lau SSY (2019) Undefined: a review of net zero energy buildings in hot and humid climates: experience learned from 34 case study buildings. In: Renewable and sustainable energy reviews. Elsevier
28. Besser D, Vogdt FU (2017) First steps towards low energy buildings: how far are Chilean dwellings from nearly zero-energy performances? Energy Procedia 132:81–86. https://doi.org/10.1016/J.EGYPRO.2017.09.642
29. Fong KF, Lee CK (2012) Towards net zero energy design for low-rise residential buildings in subtropical Hong Kong. Appl Energy 93:686–694. https://doi.org/10.1016/J.APENERGY.2012.01.006
30. Zhang S, Wang K, Xu W, Iyer-Raniga U, Athienitis A, Ge H, woo Cho D, Feng W, Okumiya M, Yoon G, Mazria E, Lyu Y (2021) Policy recommendations for the zero energy building promotion towards carbon neutral in Asia-Pacific Region. Energy Policy 159:112661. https://doi.org/10.1016/J.ENPOL.2021.112661
31. D'Agostino D, Tzeiranaki ST, Zangheri P, Bertoldi P (2021) Assessing nearly zero energy buildings (NZEBs) development in Europe. Energ Strat Rev 36:100680. https://doi.org/10.1016/J.ESR.2021.100680
32. Torcellini P, Pless S, Deru M, Griffith B, Long N, Judkoff R (2006) Lessons learned from case studies of six high-performance buildings
33. Paul Torcellini BA, Ashrae M, Crawley DB (2006) The following article was published in ASHRAE J
34. Jaysawal RK, Chakraborty S, Elangovan D, Padmanaban S (2022) Concept of net zero energy buildings (NZEB)—a literature review. Clean Eng Technol 11. https://doi.org/10.1016/j.clet.2022.100582
35. Energy Performance of Buildings Directive EU/2024/1275. https://eur-lex.europa.eu/legal-content/EN/TXT/HTML/?uri=OJ:L_202401275
36. Wilson A (2023) Energy performance of buildings directive. Members' Research Service PE 739:377
37. Charron R, A review of low and net-zero energy solar home initiatives
38. Towards net zero energy solar buildings planned outcomes (2010)
39. Besant RW, Dumont RS, Schoenau G The Saskatchewan conservation house: some preliminary performance results (1979)
40. Australian Building Codes Board | ABCB, https://www.abcb.gov.au/. Last accessed 15 July 2024
41. Directive 2010/31/EU of the European Parliament and of the Council of 19 May 2010 on the energy performance of buildings
42. School buildings-achieving advanced energy design guide for K−12 school buildings advanced energy design guide for K−12 school buildings achieving zero energy (2018)
43. AEDG—Advanced Energy Design Guides. https://www.ashrae.org/technical-resources/aedgs. Last accessed 15 July 2024
44. ZERO code: the future has arrived—architecture 2030. https://www.architecture2030.org/zero-code/. Last accessed 15 July 2024
45. Canadian home builders' association net zero home labelling program for new homes technical requirements-version 1.4
46. Liu Z, Zhou Q, Tian Z, He BJ, Jin G (2019) A comprehensive analysis on definitions, development, and policies of nearly zero energy buildings in China. Renew Sustain Energy Rev 114:109314. https://doi.org/10.1016/J.RSER.2019.109314

47. GB/T 51350-2019 English PDF. https://www.chinesestandard.net/PDF/English.aspx/GBT 51350-2019. Last accessed 15 July 2024
48. China Building Energy Conservation Association. https://www.cabee.org/. Last accessed 15 July 2024
49. SHASE: definitions and evaluation method of ZEB (net Zero Energy Building) (2015)
50. Design guidelines & pamphlets for planning and designing of zero energy buildings (ZEBs) in Japan sustainable open innovation initiative (2018)
51. Ministry of Land, Infrastructure and Transport. https://www.molit.go.kr/english/USR/BORD0201/m_28286/DTL.jsp?id=eng0301&cate=&mode=view&idx=2740&key=&search=&search_regdate_s=&search_regdate_e=&order=&desc=asc&srch_prc_stts=&item_num=0&search_dept_id=&search_dept_nm=&srch_usr_nm=N&srch_usr_titl=N&srch_usr_ctnt=N&srch_mng_nm=N&old_dept_nm=&search_gbn=&search_section=&source=&search1=&lcmspage=18. Last accessed 15 July 2024
52. Ministry of Construction of Russia. https://www.minstroyrf.ru/. Last accessed 15 July 2024
53. Republic of Turkey Ministry of Energy and Natural Resources. https://enerji.gov.tr/. Last accessed 15 July 2024
54. Energy Conservation Building Code (ECBC) | Bureau of Energy Efficiency, Government of India, Ministry of Power. https://beeindia.gov.in/en/energy-conservation-building-code-ecbc. Last accessed 15 July 2024
55. Normas Oficiales Mexicanas en Eficiencia Energética | Comisión Nacional para el Uso Eficiente de la Energía | Gobierno | gob.mx. https://www.gob.mx/conuee/acciones-y-programas/normas-oficiales-mexicanas-en-eficiencia-energetica-vigentes. Last accessed 15 July 2024
56. CONUEE—Promotion of energy efficiency and development of national standards—Promoción de la Eficiencia Energética y emitir normas oficiales mexicanas de la materia—Policies—IEA. https://prod.iea.org/policies/2614-conuee-promotion-of-energy-efficiency-and-development-of-national-standards-promocion-de-la-eficiencia-energetica-y-emitir-normas-oficiales-mexicanas-de-la-materia. Last accessed 15 July 2024
57. Building technical regulations (TEK17) with guidance—directorate for building quality. https://www.dibk.no/regelverk/byggteknisk-forskrift-tek17. Last accessed 15 July 2024
58. Schweizer Baustandard—Minergie, https://www.minergie.ch/de/. Last accessed 15 July 2024

Chapter 3
Site and Building Design Strategies for Net Zero Energy Buildings

Abstract Adapting designs to local climate conditions is an important consideration in achieving energy efficient structures. This chapter provides a broad look at zero energy building design strategies including climate considerations, passive design, energy efficiency, active design strategies, and on-site energy generation. Particularly, it examines the integration of passive design measures with active systems to ensure that buildings remain energy-efficient, resilient, and aligned with long-term climate neutrality goals. In addition, it discusses the importance of operational and embodied energy in building life-cycle assessments.

Keywords Nearly zero-energy buildings (nZEBs) · Energy efficiency · Energy conservation measures (ECMs) · Sustainable building design · Greenhouse gas reduction · Future climate adaptation

Abbreviations and Acronyms

ECM	Energy Conservation Measures
EPBD	Energy Performance of Buildings Directive
EU	European Union
GHG	Greenhouse Gas
GSHP	Ground Source Heat Pump
HVAC	Heating, Ventilation, and Air Conditioning
IHG	Internal Heat Gain
nZEB	Nearly Zero-Energy Building
PCM	Phase Change Material
PV	Photovoltaic
SHGC	Solar Heat Gain Coefficient
U-value	Thermal Transmittance

B. Duraković, *Zero Energy Buildings*, Green Energy and Technology,
https://doi.org/10.1007/978-3-032-16027-0_3

3.1 Introduction

Due to increasing energy consumption and climate change, the construction industry is attempting to adopt sustainable solutions that mitigate adverse impacts on the environment, as well as enhance energy efficiency in buildings. The construction industry is generally recognized as a primary contributing factor in relation to worldwide carbon emissions and energy consumption [1]. Between 2000 and 2017, building energy consumption grew by more than 20%, posing a serious threat to sustainable growth in society, economy, and the environment [2]. To mitigate this impact, renewable energy sources and high-efficiency technology are introduced. In recent years, there has been increased interest in designing and operating zero-energy buildings (ZEBs) to reduce dependence on conventional sources of energy [3]. The nearly zero energy building (nZEB) idea is specified in the EU's Energy Performance of Buildings Directive (EPBD), accepted in 2002 and then revised multiple times to match increasing amounts of energy and climate objectives. New buildings must comply with nZEB standards of the EPBD, which started in January 2021 [4]. An nZEB is characterized as a building with high energy performance, with almost zero or low energy requirements that are mainly provided by renewable energy sources, including on-site and nearby sources. EU Member States have come up with their own meanings based on the general nZEB set out in the EPBD. These definitions are climate, social and economic specific at the national, regional and/or local level [5]. This has led to a variety of definitions as well as varying aspirations. The most frequent method to define this concept consists of controlling the energy balance of a single building in a year, computed with on-site energy from renewable sources, and primary energy demand for heating, cooling, ventilation, bathrooms domestic water, built-in lighting or additional energy. Few definitions include higher energy consumption, for example appliances and central services [6, 7]. Previous research shows a considerable variation in compliance with nZEB regulation between the different areas of Europe. For example, Northern European countries with central heating and geothermal energy, like Finland, Sweden, Estonia and Norway, define nZEB differently Many buildings in countries such as Finland and Norway can reach nZEB standards if they use passive energy management techniques alone. Conversely, in Sweden and Estonia, buildings need to have on-site green energy resources, such as photovoltaic (PV) systems, to comply with these regulations [7].

Requested focus areas include, but are not limited to, a reduction in emissions from production of materials and the transition of energy supply (e.g. electricity, central heating, etc.) from non-renewable to renewable energy sources. Buildings must demonstrate how they achieve carbon neutrality by 2050 and align with zero-energy building standards. This objective has been advanced by the energy efficiency interventions available for the reduction of heating, cooling and hot water [8]. To effectively adapt to climate change and meet energy performance goals, Zero-Energy Buildings (ZEBs) must incorporate a range of integrated design strategies. These strategies address climate responsiveness, passive design, energy efficiency, and on-site energy generation. Figure 3.1 represents some of the common ZEB design strategies discussed in this chapter.

Climate considerations, such as optimal site orientation, local climate analysis, seasonal solar angles, and wind patterns, shape building design for natural heating, cooling, and lighting. Passive strategies, including envelope orientation, geometric design, and thermal energy storage (e.g., Phase Change Materials), minimize the need for mechanical systems. Energy efficiency is enhanced through optimized HVAC, lighting, water heating, appliances, and building automation. On-site energy generation, using PV systems, wind turbines, solar collectors, and Combined Cooling, Heating, and Power (CHP), ensures that buildings meet their energy needs sustainably. Together, these strategies create buildings that are energy-efficient, climate-adaptive, and capable of achieving net-zero energy consumption.

Energy efficiency measures are defined as strategies, technologies, or reformation efforts that aim to reduce a building energy consumption with appropriate maintenance of indoor thermal comfort [9, 10]. Energy-efficient procedures can be passive design approaches (such as thermal insulation of the structure envelope as well as high-efficiency home windows) or active design approaches, as well as renewable techniques, such as energy-efficient heating, ventilation, and air conditioning (HVAC) systems and renewable resource sources, such as building-integrated Photovoltaic panels. The research also demonstrates the importance of embodied energy when estimating the full life cycle of a building not only its operating energy. Embodied energy refers to the energy used to produce things away from the site, deliver them, build the building, keep it maintained, repair it, and after life periods finish, bring it down or recover it. The reason is that, ultimately the decision must be made, on one side, the less energy will be consumed in the lifetime of a building (for example, to heat and to cool), which would need to be built with materials or

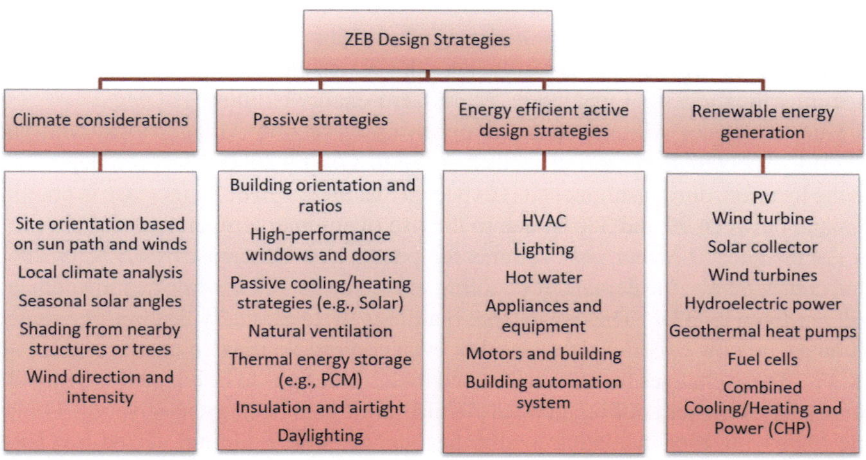

Fig. 3.1 ZEB design strategies

systems with higher embodied energy. However, the use of low-energy, stored potential materials may imply that more energy is being consumed during the use of the building. Understanding both forms of energy with the entire life cycle of the building is key to designing truly sustainable, energy-efficient buildings. Simply minimizing operational energy without thinking how to install passive and active measures for energy efficiency, as well as renewable energy, affects the overall lifecycle of the building and is shortsighted move. It would be very ironic if effort in reducing operational energy resulted in increased embodied energy [11]. Thus, to remain true to the long-term objective of climate neutrality by 2050, one must follow the cost–benefit of dependence on embodied and operational energy. This approach can allow achieving life-cycle net zero energy/GHG emissions of the buildings relative to embodied and operational energy through the use of renewables.

3.2 Climate Considerations

The world views climate change as one of the greatest threats of this century. These effects are severe and complex impacting climate, health, and the economy. Because of these factors fighting climate change must be done rapidly [12]. Summer heat waves with temperatures above 35 °C have become more frequent and severe over the last ten years. Over this period, extreme weather events such as extreme temperatures, rainfall, humidity, wind, and sunlight reaching the Earth have all changed. Temperatures in summer, not being lower than 35 °C, have been observed more often and have been more severe throughout this decade. During this time, extreme weather events, the hottest temperatures, the heaviest rainfall, floods, humidity, wind, and sunlight hitting the Earth have all changed [13]. Because of these climate change impacts, people increasingly seek protection and comfort inside buildings [14]. This affects energy consumptions on heating and cooling causing more greenhouse gas emissions [15]. Climate change influences the design and construction of buildings, including the process of making it "nearly zero-energy building".

Due to warmer winters, some European cities such as Milan, Stockholm, and Madrid, expect to drop heating needs from 38 to 57% by 2060. However, depending on the location, during summers it is expected that there will be an increase in cooling demand between 99 and 380%, due to the rise of summer temperatures.

For instance, in Milan, peak summer temperatures will increase by 4.3 °C by 2060 and, consequently, the demand for cooling will go up by 107% while heating demand will go down by 37%. This means that buildings in traditionally heating-dominated climates are now experiencing mixed or even cooling-dominated energy needs [14, 16, 17]. These changes then show the need for ZEB designs to be adapted to deal with the reality of higher cooling demands whilst minimizing energy use overall. Things like improving roof insulation to stop heat entering the building, glazed low solar gain windows that limit the amount of heat transmitted through glass, and ensuring that the building envelope is as airtight as possible to prevent unwanted heat access during hot weather, are some of the examples of strategies when designing. These measures reduce the need for air conditioning and keep indoors comfortable during extreme heat.

To summarize, ZEBs must adapt to climate change and the changing net balance of heating and cooling needs by implementing strategies that account for overheating risks alongside continued plans for energy efficiency. Strategies such as improved insulation, advanced window designs, and accurate and incorporated metrics for climate data should be integrated into building designs. That adaptation is essential for lowering energy demand by keeping occupant comfort and functionality as temperatures rise in the coming decades [17, 18].

3.3 Passive Design Strategies

Passive design strategies are energy conservation measures (ECM) that do not require electricity or mechanical system [19]. Well planned and designed buildings are a primary line of defense against global warming, minimizing summer heat absorption and dissipating less heat in winter [20]. Table 3.1 summarizes some major strategies of Passive Design and its benefits in practice. Passive concepts can be used to minimize energy consumption and improve living quality. However, it focuses on indoor heating also. Each one of such design ideas will be elaborated in the following paragraphs. The table also provides a fast overview of these strategies effect on building performance and environmental outcomes.

The passive ECMs considered are thermal insulation of building envelopes, specific use of materials, green envelopes, shading, thermal mass and PCM, ventilation, window design, building design interventions like orientation and window to wall ratio, and airtightness. Another type of passive measure tackles how occupants behave. There is research to establish individual linear failure to evaluate the significance of various passive ECM for single family detached buildings [36]. The outcomes highlighted the changing role of passive ECMs in connection with climate change. As it turns out, the most important conduit passage of heating loads was that of the impermeable envelope U-value. In contrast, the window-to-floor ratio was found equally important to cooling loads in current and future conditions.

3.3.1 Passive Solar Heating/Cooling

Passive solar heating and cooling strategies utilize natural sunlight, wind, and thermal mass to regulate indoor temperatures in order to reduce the need for mechanical heating and cooling. These strategies include techniques like solar walls [37], earth air heat exchangers; and natural ventilation systems, such as wind towers, and solar energy collectors [24].

Table 3.1 Passive design strategies and benefits

Passive strategies	Description	Benefits	Ref
Thermal insulation	Insulating walls, roofs, and floors to minimize heat transfer	Reduces heating and cooling energy by 15.5–19.9%; mitigates extreme temperatures	[21, 22]
Passive solar heating/ cooling	Using building orientation, thermal mass, and shading to regulate temperature	Decreases heating and cooling loads; improves occupant comfort; reduces energy costs	[23–25]
Natural ventilation	Designing openings and airflow paths to promote passive air movement	Enhances indoor air quality; reduces reliance on mechanical cooling; lowers energy consumption	[26–28]
Thermal energy storage	Integrating materials that absorb and release heat to maintain stable indoor temperatures	Enhances thermal comfort; reduces peak energy demand; lowers heating and cooling costs	[29]
Cool Roofs	Using reflective materials to reduce heat absorption	Cuts cooling energy by 14–22% in warm climates; mitigates urban heat islands	[30]
Green envelope	Incorporating vegetation on roofs and walls to reduce heat transfer	Reduces cooling demand by 2–50%; enhances air quality and reduces greenhouse gas emissions	[31, 32]
Shading and solar control	Installing external shading systems to limit solar heat gain	Decrease cooling demand by up to 62.7%; reduces interior overheating	[33]
High-performance windows and doors	Optimizing window orientation, glazing, and shading to control solar heat gain and loss	Improves daylighting; reduces cooling loads in hot climates and heating in cold climates	[34]
Building orientation	Positioning the building for optimal sunlight and natural ventilation	Maximizes solar gain in winter and reduces overheating in summer	[35]

Passive strategies contribute to cost reduction, energy savings and maintaining thermal comfort. The strategic placement of windows and shading devices can maximize solar gain during winter and minimize it during summer, reducing heating and cooling demands. The use of materials with high thermal mass allows buildings to store heat during the day and release it at night, reducing interior temperature fluctuation. These approaches are particularly effective in regions with significant diurnal temperature variations [25].

3.3.2 Natural Ventilation

Natural ventilation utilizes natural buoyancy-driven stack effect to promote air circulation within buildings, reducing reliance on mechanical cooling systems. Key techniques include cross ventilation, which involves placing openings on opposite sides of a building to facilitate airflow, and stack ventilation, where warm indoor air rises and exits through higher openings, drawing cooler air in through lower openings. The effectiveness of these strategies varies with climate conditions and could maintain indoor comfort year-round in warmer climates [27]. For the implementation of natural ventilation, it requires proper building orientation, window placement, and local wind patterns [28].

3.3.3 Solar Reflections

As the need for cooling grows, external opaque surfaces become more crucial [36]. Use materials that reflect the sun in the building envelope to keep the surface cooler than with a typical envelope. This will lower the building cooling expenditure. It is mentioned that highly reflective roofs will become more common to lower the need for cooling [30]. Based on the weather changes happening now and through the future, it is expected that cool roofs in warm and subtropical places could cut energy use by 14% and 22%, respectively [38]. In cool and moderately mild climates, this ECM initially made total energy consumption higher (a heating cost in the winter) in the way things were set up. Researchers have looked at how cool and green roofs affected the internal environment of an office building. They found that, cool roofs were better than green roofs at keeping buildings from getting too hot in the summer [39]. In Montreal, where it is very cold, Hosseini et al. [40] discovered that roofs with more sun reflection lower cooling energy and variation, which makes the design more resistant to global warming. Using building materials that do not absorb much sunlight could help save energy, especially in hot places.

3.3.4 Green Envelope

Green building envelopes, such as green roofs and green facades, have potential to lower the urban heat island and in the same way improve the air quality by using plants to stop heat from entering the building structure and help cool the envelope through evaporation. Green walls cut down on energy use and greenhouse gas emissions more than green roofs [31]. Chan and Chow [32] found that the green roof decreased the amount of energy used for air cooling (A/C) by 2.4–10% from 2011 to 2030 and 2046 to 2065, compared to Hong Kong levels. It is found that adding green roofs to old buildings might lower the energy used for heating and cooling and keep naturally aired buildings from getting too hot [41]. In 2020, 2050, and 2080, weather scenarios and four building renovation in Qatar, green walls and roofs may save 2–4% of its energy [42]. Green roofs contributed to between 20 and 50% of energy savings in Esch-sur-Alzette (Luxembourg) and 3% to 15% in Palermo (Italy) as well as upper floor ceiling temperature reduction from 2° to 5° [43]. Getting enough water in the summer limits how well it works because it makes surface latent heat loss more severe. Peak temperatures above 40 °C may happen more often in Athens and other European cities [44]. Most plants can live in temperatures between 0 and 40 °C, so it is important to think about how well green roofs will perform in cold weather.

3.3.5 Shading and Solar Control

Solar shading devices lower the solar heat gain through the windows and the amount of cooling [45]. Huang and Hwang [46] found that shading devices in Taiwan in 2020, 2050, and 2080 would balance out 65.5, 37.5, and 27.7% of the energy used to cool the air. According to Silva et al. [33], night air reduced high cooling energy by 3.5–10.9% and shade cut by 46.4–62.7%. Louvered shading cut down on temperature rise hours more than outdoor insulation, high brightness, and present mass in four types of domestic buildings [47]. Shade on the outside of the building prevents solar heat gain. However, this study was done in Burkina Faso, West Africa, which is humid and has high average temperatures. Therefore, cooling issues are more challenging than heating ones. Urban green spaces like parks, grasslands, water bodies, and trees can be more cost-effective options, better for the environment than shade systems that are directly integrated into buildings [48]. Blue-green features in cities can shade and cool down nearby buildings, lowering the effects of urban heat island and temperature changes caused by climate change. Their effectiveness changes depending on the size and number of plants, the weather, the time of day, and the region.

3.3.6 Building Orientation and Landscape Design

The orientation of a building refers to the direction that it faces. It plays an important role in how much sunlight, wind, and natural ventilation it receives. Therefore, it is a critical design consideration for energy efficiency. Correct orientation allows for maximum solar gain in the colder months while not creating overheating in the warmer months. In cold climates, south-facing facades provide maximum solar gain, reducing the need for mechanical heating systems. However, in hot areas, limiting direct sun falling on east and west-facing walls through building orientation or shading devices can yield extensive reductions in cooling loads. It is recommended to take the local climatic conditions into account for orientation decisions. For example, while buildings in the tropics benefit from minimizing east–west exposure to reduce solar heat gain, buildings in the temperate zone can improve daylighting and passive heating of spaces by orienting the primary living spaces toward the sun. The planting of vegetation and the inclusion of reflective surfaces around the building also address these objectives by minimizing urban heat island impacts [49]. It was found that south-facing building was best and north-facing building was worst in conserving energy [35]. Yan et al. [45] identified a south-easterly orientation as better than the base case for future scenarios. Most building orientation adjustments can only work for new buildings or complete reconstructions, and when it comes to old buildings, a site can be a roadblock to changing the orientation as mentioned, in addition to urban planning rules and the costs involved. This means developers must take climate change into account when designing new buildings, as orientation choices will affect their energy performance for several decades to come.

3.3.7 High-Performance Windows and Doors

Window and doors design attributes are the second most common ECMs and facilitate buildings exchanging heat and sunlight. Certain glazing technologies can increase thermal comfort hours by 14.2% contributing energy demand reduction [50]. Across the study of a humid and hot environment, the solar heat gain coefficient (SHGC) was found to be the most influential design parameter [38]. In the Hong Kong sensitivity analysis, Liu et al. [34], thus SHGC was more influential than the wall solar absorptance or the shading device factor. Window SHGC was found to enhance efficiency and global warming resistance [51]. In Sweden, the top retrofitting measure among five national factors for minimizing heating demand was energy-efficient windows and envelope insulation. Window design, however, needs to be adapted for place and climate. Low window U-values are useful for many climates. Low U-value glass is better suited for hot climate areas, such as those found in Hong Kong, Taiwan, and the UAE, where cooling should dominate, and heating needs are insignificant [21].

3.3.8 Building Envelopes and Air Tightness

As the primary barrier against external environmental influences, insulating building envelopes is a vital strategy for energy conservation to combat climate change. According to Radhi's research [21], an increase in insulation can enable a reduction of cooling energy consumption by 15.5–19.9% over different climatic scenarios. Pajek et al. [22] case that the most impactful ECM is to reduce the envelope U-value, is less effective in warmer climates. This difference is partially a result of Montenegro's status as one of the least energy-efficient countries in Europe, with generally poor thermal insulation [22]. Conversely, Waddicor et al. [52] which reported a clear lack of effect in their study area that buildings already had 10 cm of wall insulation. This variation in outcomes highlights a few key determinants of ECM effectiveness: (1) regional climate conditions that influence whether energy savings should come from heating (in colder climates) or cooling (in warmer climates) (2) existing building codes, if a region already has generous insulation regulations, those buildings will receive less of a return from retrofitting than those without (3) baseline insulation levels, so poorly insulated buildings or those built pre-thermal regulations will continue to gain from added insulation [53].

Research by Rodrigues et al. [31], found that there is no singularly outstanding thermal transmittance (U-value) value for modern day renovations but is based on atmospheric conditions. This may be either below or above a certain standard value for the ideal thermal transmittance, as the heating needs can be low or high. The complex interaction between ambient temperature and building capacity to release heat could lead to a larger overall optimal thermal transmittance. As average temperatures increase in some regions, outdoor conditions can reach indoor thermal comfort setpoints. In areas with high cooling loads, though, optimum thermal transmittance can be low, since average temperature will exceed cooling thermostat set points and daily minimums will approach that of heating thermostat set points in both current and expected-to-become climates. Thermal transmittance must be minimum in such cases, to restrict heat absorption through the building envelope [31].

An example of this would be Singapore, where emphasized that envelope insulation upgrades are primary means of residential building thermal performance improvement [45].

In addition, it is also important to note that ideal values for thermal transmittance can vary significantly between different building classes even for the same envelope elements, especially perimeters such as roofs and ceilings which are exposed to solar radiation. Although improved insulation reduces heating energy demand in general, hot periods with the associated added risk of overheating and increased cooling energy demands have also risen. Jafarpur et al. [54] shown that increasing insulation levels together with the use of triple-glazed windows can negatively impact energy consumption in summer due to increased cooling demand for climates such as Vancouver. In contrast, Quebec case showed that added insulation is still valuable in colder climates. Envelope insulation alone may be less significant benefit in future climates due to global warming [55].

Although improved airtightness can reduce the heating needs associated with cold snaps in the absence of effective ventilation strategies, such measures may promote overheating during summer heat extremes [20]. This situation further highlights the need for balance between high levels of insulation and using appropriate passive cooling strategies, such as natural ventilation, night cooling and shading.

In conclusion, this now implies that when applied in conjunction with purpose performance building control measures both responsive to interdisciplinary sustainability pathways, while at the same time feasible with respect to yearly principles, they have prospects towards addressing desired energy savings while physical-handlings like incidental heat losses or gains can be adequately queued for providing thermal comfort.

To understand infiltration rates and airtightness, it is required to understand what these terms mean in the context of the ability of air to pass between outside and inside. Supporting sensible building design in nearly-zero energy buildings, especially in dynamic climates, reducing infiltration as much as possible is critical as this has massive energy savings potential, nearly as much as insulation [56]. Liu et al. [57] suggested that future climatic conditions should guide improvements in building airtightness. In regions such as Hong Kong, where residential buildings tend to have a low standard of airtightness, a higher level of infiltration might be a cost-effective alternative. Shen et al. [58] ranked infiltration rates as the predominant ECM affecting energy consumption for several building types. One key consideration is that an airtight building, which would require less energy consumption to stay warm, could make retrofitting less economically feasible over time.

In addition, the performance of airtightness and insulation measures can vary substantially depending on the operational modes of buildings (full air-conditioning, free-running, mixed-mode), including the associated heating and/or cooling loads. Locations that are hot and dry and with climate change use significantly more air conditioning, enhanced low level airtightness can still be a positive step, even for air-conditioned buildings. Free running structures in temperate climates can often benefit from higher levels of infiltration in the passive cooling strategies that keep the indoor temperatures moderated. Mixed-mode buildings need to tune their airtightness in the cooling operation of air conditioning and heat dissipating in the case of natural ventilation. Also, the new complexities from climate change will change environmental conditions, complicating this balance even more. Thus, new buildings' airtightness needs to be carefully treated considering the influence of local climate from a climate change perspective and the operational strategy to reach an optimal compromise among energy efficiency, indoor air quality and thermal comfort.

3.4 Energy Efficient Active Design Strategies

While passive design strategies reduce the total amount of energy that a building needs, active systems focus on increasing the efficiency of the building energy systems. This includes the usage of modern technologies in necessary systems for the building like heating, cooling, lighting, and electrical systems for energy optimization purposes. Recent advancements in these smart technologies have allowed buildings to contribute to energy savings, without sacrificing costs to comfort and productivity for residents of the building. This progress is primarily related to sustainability and includes overall operational efficiency. In ZEB design, tackling a number of challenges is required in order to ensure energy efficiency and sustainability. These challenges range from the rise in cooling loads caused by climate change to the energy embodied in materials, and intermittent renewable energies. Table 3.2 shows some of these challenges when designing ZEBs and solutions to them.

The main task when designing ZEBs is to meet the increased need for cooling brought on by global warming without air conditioning. To do that, today's engineers must come up with new ways of doing things like using advanced shaded devices, cool roofs and envelopes around the structure that are airtight and fully insulated. Moreover, building materials contribute a significant share to the first apparent energy-only

Table 3.2 Challenges and solutions for ZEB design

Challenge	Impact	Research direction	References
Increasing cooling demand	Higher cooling loads due to rising temperatures	Passive design measures like cool roofs, shading devices, and airtight envelopes	[12]
Embodied energy in materials	High carbon footprint from material production and transport	Use of low-carbon and recyclable materials; focus on lifecycle energy optimization	[21]
Variability in ZEB definitions	Inconsistent standards across regions; compliance challenges	Harmonization of nZEB definitions based on climate and regional factors	[4, 6, 7]
Renewable energy storage and reliability	Intermittency of solar and wind energy sources	Integration of complementary systems (e.g., geothermal, battery storage) and grid-interactive technologies	[59–61]

assessment of buildings, dramatically expanding their carbon footprint. This argues for low carbon materials and recycling in construction. One more difficulty lies in the differing definitions of what exactly it implies to be a zero-energy building from area to area. This leads to problems in compliance and application. It is essential to consolidate these definitions in line with local climatic and social conditions. In addition, the variability of renewable energies such as solar and wind presents continuity and reliability issues. These issues show how challenging ZEB design can be and requires the need for multiple approaches in achieving sustainability goals.

3.4.1 HVAC Systems

There is different energy sources used for heating, some systems are using natural gas and other systems are consuming electricity to heat the interior of the building. There are different investigations of various configurations of HVAC systems serving as active energy conservation measures, which include district heating (DH), electric heating (EH), and heat pumps. Rahif et al. found that an electricity-based systems reduced greenhouse gas emissions by 15–27% more than a fossil fuel-based strategy and used about 6–13% less raw energy. In the same way, Tarroja et al. [62] showed that using electric heaters would cut greenhouse gas emissions by 30–40%. These studies stressed how important it is to make the energy sector less carbon-based and to electrify the building sector. The main point is that the rise in electricity usage in the future would have almost no effect on greenhouse gas emissions because renewable energy sources would be the main source of electricity production in 10 years. But there are problems that need to be addressed, such as the cost, the return effect, and the need for more grid resource capability. Heating loads can still be powered by electricity when PV energy is low like during the time of winter. For instance, they speculated the high price of natural gas and electricity has made it increasingly difficult for Canada to switch to electric heating [63]. Heating, ventilation and air conditioning (HVAC), which controls the thermal comfort in a full range of buildings, consumes a lot of energy in any kind of building, especially those with severe temperature swings. One way to change this pattern in zero-energy buildings is by making the systems more efficient. The introduction of smart HVAC system integration with building management systems allows building energy use to be continuously monitored and controlled in real time. These systems can analyze patterns such as occupancy and weather predictions, meaning that energy is only consumed when it actually needs to be used. These types of controllers can reduce residential and commercial building mechanical system energy consumption while maintaining satisfaction.

3.4.2 Electrical Systems

The efficiency of air conditioning appliances has increased up. For example, the efficiency of split-type air conditioners has gone higher by 3% every year in the Asian, European, American, and Australian markets over the last 15 years [64]. In the future, this development is likely to keep happening in air-conditioning systems. So, as an active Energy conservation measure, the effect of energy-efficient chillers on future heat performance has been investigated. A big energy savings of about 16 kWh/m2 in total primary energy was made by raising the chiller's coefficient of performance from 3.0 to 6.0 [52]. Barea et al. [65] investigated how chiller coefficient of performance has changed over time and predicted that it would rise from 2.3 in 2020 to 3.5 in 2075 and then to 4.4 in 2099. According to the study, Argentina's cooling energy use would drop even more than it is now. It is important to keep in mind, though, that while making air conditioners more efficient is good, putting them in a lot of places that did not have them before could make a big difference in the amount of energy that is needed. Researchers have found that putting a lot of these devices in places that are getting warmer could cause big jumps in energy use, which gets worse as the population grows [66].

3.4.3 Lighting

Another source of heat in a house that makes it warmer is Internal heat gain (IHG). For IHG, ECM does a review of the energy use of electrical devices that make heat when they are turned on or started. Because IHG has such a big effect on season rising temperatures, more and more buildings that are efficiently insulated use energy-efficient lights and electronics. Some references indicate how reducing internal heat loads could help save energy. In a study Li et al. [67] found that lowering the brightness of lights in structures that are not residential, except in very cold places, is a great way to save energy. Berger et al. [68] also proved that efficient lighting and equipment that worked better had a greater effect on how much cooling was needed in most office buildings. Similar results were found by Daly et al., who found that the intensity of devices could make a big difference in how much energy it uses. High power intensity was linked to 36% more energy use, while low intensity was linked to 20% less energy use [69]. By changing the light in each office, it reduced energy usage by 20.1–25.4%. This reduces heat gain as much as possible to save energy and reduce the office building's greenhouse gas emissions [70]. Mitigating the heat from the lighting and electric devices will result in reducing the HVAC load as the less heat will left for cooling and heating [71].

3.5 Renewable Energy Generation

Here it is discussed how innovative energy systems can help cut down on traditional power sources and how adding green energy sources to buildings is becoming more important. It is important to make remarks on renewable energy as the energy conservation measure, even though using electricity from renewable sources does not directly save energy like fossil fuels do. This is because using electricity from renewable sources lowers primary energy use and carbon emissions. And so, green energy should be seen as a bigger ECM plan. This includes technologies like wind turbines, solar heating systems, building-integrated PV systems, and geothermal systems. The distribution of various renewable forms of energy used in a small volume building is illustrated in the Fig. 3.2. Photovoltaic systems have developed as the overwhelming proportion of renewable systems due to their scalability, descending expenses and being capable of on-site electricity production. They accounted for 50% of total energy generation in 2023. Geothermal systems (25%) use ground-source heat pumps to supply heating and cooling in residential buildings. In terms of HVAC energy use, geothermal systems surpass standard air-source heat pumps. There are a significant advancements in wind turbine technologies [72]. Wind turbines supply 15%, as a steady energy source where there are reliable wind patterns to smooth out short periods without light solar. And 10% of contributions are from solar hot water systems which augment thermal energy for both domestic water heating and space heating. This diversity of advantages suits the different kinds of renewable energy technologies needed for ZEBs in conserving energy and environmentally benign houses. The chart shows that to meet the demand for energy, reliability and performance various renewables are necessary.

Due to its flexible nature, scalability, and decreasing costs, solar energy stands as the most used renewable resource in ZEBs. This is most often accomplished through PV panels that generate on-site energy by absorbing sunlight to convert it into usable energy for heating, ventilation, air conditioning (HVAC), or electrical loads [74]. Improved efficiencies in solar panels and more effective energy storage systems mean that buildings can consume all the energy they generate or more energy than they need. There are two studies that look at how changing climate affects PV systems that are built into houses. In 2060 climate scenarios, D'Agostino et al. [59]

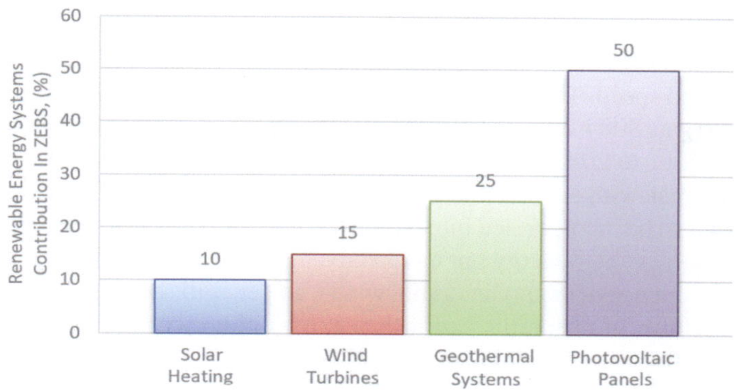

Fig. 3.2 Renewable energy systems contribution in ZEBs. *Source*: Author based on [73]

found that building-integrated PV power went up by 3–20% in European towns. In Australia [75] and the UK [61], the same trend can be seen. Building-integrated PV power is further complicated due to amount of energy the building needs. According to Robert and Kummert [13], the summer energy surplus has been going down over time. The amount of surplus energy in forecasts for later years was also negligible due to increased cooling needs [59]. In fact, the energy shortage could happen in July 2050 because a significant increase in cooling demand is expected. Low sun angles and short winter days are not significant contributors to heating as they are not adequate for typical houses [59]. In other words, the need for heat must still be considered when ZEBs during the design process. Geothermal energy provides constant temperatures from below the Earth's surface for the purpose of heating, cooling, as well as electricity generation. A common geothermal technology used in ZEBs is ground-source heat pumps (GSHPs), which transfer heat between the building and the ground. When a building does not require heating, GSHPs transfer heat from the building to the ground, and vice versa. In an area with extreme temperature swings, they provide reassurance of a source of energy that operates well. GSHPs achieve efficiencies of 300% or 400%, which means geothermal systems significantly lower HVAC energy consumption [60]. Geothermal technologies support solar and wind energy systems so that they can improve buildings' energy performance and enhance resilience. Therefore it is recommended to use renewable electricity and green thermal energy sources to heat and cool buildings [76].

Recently, heat pumps have become a popular option compared to boilers and other solid fuel systems because they are three to four times more efficient [77]. In the winter, ground source heat pumps use the relatively stable temperature of lower ground to get heat from the ground [78]. A study by Jylhä et al. [79] looked at three different cases and compared them: (1) mechanical cooling with electricity and heat; (2) mechanical cooling with district heating; and (3) ground source heat pumps and underground free cooling. Luo and Oyedele [61] looked at a similar topic and discussed a new trend in solar heater thermal power output. They found that it could increase by about 30 kWh per year between 2021 and 2040 and 2061 and

2080, compared to only 70 kWh per year from 1981 to 2000 to 2021–2040. In wind alternatives, the amount of energy generated by wind turbines is closely linked to changes in wind speed. The average amount of wind power generated each year from 1981 to 2000 was 800 kWh, which is more than the averages of 650 kWh from 2021 to 2040 and 2061 to 2080 [61]. Wind is a renewable energy source, and it becomes abundant in areas with stable wind patterns. Vertical-axis and horizontal-axis wind turbines can also be incorporated into buildings or placed on-site to create electricity. Not as common as solar energy but has an advantage with rural, coastal areas where there are vast open spaces and favorable winds, as you will see when looking at one of these structures in operation. For instance, a building with PV panels and wind turbines would ensure a better energy supply. Wind turbines can generate electricity during cloudy days and at night when solar generation drops, supplying the energy needs consistently [80].

3.6 Conclusions

Adapting designs to local climate conditions is an important consideration in achieving energy efficient structures. In this regard, it is required to use innovative materials and solutions to create buildings that utilize the most of natural heating/cooling, and ventilation. These localized passive techniques can reduce the need for energy-intensive solutions contributing to sustainability. Adapted designs to local climates can enhance resilience and contributes to the broader goal in meeting international climate targets.

References

1. Hanif I (2018) Impact of fossil fuels energy consumption, energy policies, and urban sprawl on carbon emissions in East Asia and the Pacific: a panel investigation. Energ Strat Rev 21:16–24. https://doi.org/10.1016/J.ESR.2018.04.006
2. Lu M, Sun Y, Ma Z (2024) Multi-objective design optimization of multiple energy systems in net/nearly zero energy buildings under uncertainty correlations. Appl Energy 370:123620. https://doi.org/10.1016/J.APENERGY.2024.123620
3. Saleh O, Jomah M, Ahmed A A, Alsharif A, Najway A (2024) Towards net zero energy buildings for sustainability, African J Adv Appl Sci 3(3):228–234
4. Hughes C, Molloy G (2022) New Nearly Zero Energy Buildings (NZEB), Sustainable Energy Authority of Ireland (SEAI), available at: https://cdn.ca-epbd.eu/epbd/Media/638373594343675540/CT1-New-Nearly-Zero-Energy-Buildings-NZEB--Status-in-2022.pdf?utm_source=chatgpt.com
5. Menezes AC, Cripps A, Bouchlaghem D, Buswell R (2012) Predicted vs. actual energy performance of non-domestic buildings: using post-occupancy evaluation data to reduce the performance gap. Appl Energy 97:355–364. https://doi.org/10.1016/J.APENERGY.2011.11.075

6. Simson R, Thomsen KE, Wittchen KB, Kurnitski J (2021) A comparative analysis of NZEB energy performance requirements for residential buildings in Denmark, Estonia and Finland. In: E3S web of conferences. EDP Sciences. https://doi.org/10.1051/e3sconf/202124614001
7. Reda F, Fatima Z (2019) Northern European nearly zero energy building concepts for apartment buildings using integrated solar technologies and dynamic occupancy profile: focus on Finland and other Northern European countries. Appl Energy 237:598–617. https://doi.org/10.1016/J.APENERGY.2019.01.029
8. Shadram F, Mukkavaara J (2024) Investigating the potential for realizing life cycle net-zero energy buildings in Europe using multi-objective optimization. Energy Rep 12:5648–5670. https://doi.org/10.1016/j.egyr.2024.11.042
9. Nebrida JA, Gomba FE (2023) Sustainable construction strategies for building construction projects in the Kingdom of Bahrain: a model. Sustain Eng Innov 5:31–47. https://doi.org/10.37868/SEI.V5I1.ID193
10. Nematchoua MK, Orosa JA, Reiter S (2019) Life cycle assessment of two sustainable and old neighbourhoods affected by climate change in one city in Belgium: a review. Environ Impact Assess Rev 78:106282. https://doi.org/10.1016/J.EIAR.2019.106282
11. Asdrubali F, Grazieschi G (2020) Life cycle assessment of energy efficient buildings. Energy Rep 6:270–285. https://doi.org/10.1016/j.egyr.2020.11.144
12. Santamouris M (2016) Innovating to zero the building sector in Europe: minimising the energy consumption, eradication of the energy poverty and mitigating the local climate change. Sol Energy 128:61–94. https://doi.org/10.1016/J.SOLENER.2016.01.021
13. Robert A, Kummert M (2012) Designing net-zero energy buildings for the future climate, not for the past. Build Environ 55:150–158. https://doi.org/10.1016/J.BUILDENV.2011.12.014
14. Guan L (2009) Preparation of future weather data to study the impact of climate change on buildings. Build Environ 44:793–800. https://doi.org/10.1016/J.BUILDENV.2008.05.021
15. Gaterell MR, McEvoy ME (2005) The impact of climate change uncertainties on the performance of energy efficiency measures applied to dwellings. Energy Build 37:982–995. https://doi.org/10.1016/J.ENBUILD.2004.12.015
16. Manning M, Chen Z, Marquis M (2007) Climate change 2007: the physical science basis. In: Working group I contribution to the fourth assessment report of the IPCC
17. D'Agostino D, Parker D, Epifani I, Crawley D, Lawrie L (2022) How will future climate impact the design and performance of nearly zero energy buildings (NZEBs)? Energy 240. https://doi.org/10.1016/j.energy.2021.122479
18. Moazami A, Nik VM, Carlucci S, Geving S (2019) Impacts of future weather data typology on building energy performance—investigating long-term patterns of climate change and extreme weather conditions. Appl Energy 238:696–720. https://doi.org/10.1016/j.apenergy.2019.01.085
19. Balali A, Yunusa-Kaltungo A, Edwards R (2023) A systematic review of passive energy consumption optimisation strategy selection for buildings through multiple criteria decision-making techniques. Renew Sustain Energy Rev 171:113013. https://doi.org/10.1016/J.RSER.2022.113013
20. Hong T, Malik J, Krelling A, O'Brien W, Sun K, Lamberts R, Wei M (2023) Ten questions concerning thermal resilience of buildings and occupants for climate adaptation. Build Environ 244:110806. https://doi.org/10.1016/J.BUILDENV.2023.110806
21. Radhi H (2009) Evaluating the potential impact of global warming on the UAE residential buildings—a contribution to reduce the CO_2 emissions. Build Environ 44:2451–2462. https://doi.org/10.1016/J.BUILDENV.2009.04.006
22. Pajek L, Jevrić M, Ćipranić I, Košir M (2023) A multi-aspect approach to energy retrofitting under global warming: a case of a multi-apartment building in Montenegro. J Build Eng 63:105462. https://doi.org/10.1016/J.JOBE.2022.105462
23. Duraković B (2020) Passive solar heating/cooling strategies. In: PCM-based building envelope systems: innovative energy solutions for passive design, pp 39–62. Springer Nature. https://doi.org/10.1007/978-3-030-38335-0_3

24. Gupta N, Tiwari GN (2016) Review of passive heating/cooling systems of buildings. Energy Sci Eng 4:305–333. https://doi.org/10.1002/ESE3.129
25. Uwa JN, Pourvahidi P (2024) An overview of the effects of passive solar strategies on thermal comfort. J Mediterranean Cities 4. https://doi.org/10.38027/MEDITERRANEAN-CITIES_VOL4NO1_2
26. Zhang M, Han W, He Y, Xiong J, Zhang Y (2024) Natural ventilation for cooling energy saving: typical case of public building design optimization in Guangzhou, China. Appl Sci 14:610. https://doi.org/10.3390/APP14020610
27. Tan Z, Deng X (2017) Assessment of natural ventilation potential for residential buildings across different climate zones in Australia. Atmosphere 8:177. https://doi.org/10.3390/ATMOS8090177
28. Yun GY, Steemers K, Baker N (2008) Natural ventilation in practice: linking facade design, thermal performance, occupant perception and control. Build Res Inf 36:608–624. https://doi.org/10.1080/09613210802417241
29. Simonsen G, Ravotti R, O'Neill P, Stamatiou A (2023) Biobased phase change materials in energy storage and thermal management technologies. Renew Sustain Energy Rev 184:113546. https://doi.org/10.1016/J.RSER.2023.113546
30. Karimpour M, Belusko M, Xing K, Boland J, Bruno F (2015) Impact of climate change on the design of energy efficient residential building envelopes. Energy Build 87:142–154. https://doi.org/10.1016/j.enbuild.2014.10.064
31. Rodrigues E, Fereidani NA, Fernandes MS, Gaspar AR (2023) Climate change and ideal thermal transmittance of residential buildings in Iran. J Build Eng 74:106919. https://doi.org/10.1016/J.JOBE.2023.106919
32. Chan ALS, Chow TT (2013) Energy and economic performance of green roof system under future climatic conditions in Hong Kong. Energy Build. 64:182–198. https://doi.org/10.1016/J.ENBUILD.2013.05.015
33. Silva R, Eggimann S, Fierz L, Fiorentini M, Orehounig K, Baldini L (2022) Opportunities for passive cooling to mitigate the impact of climate change in Switzerland. Build Environ 208:108574. https://doi.org/10.1016/J.BUILDENV.2021.108574
34. Liu S, Kwok YT, Lau KKL, Ouyang W, Ng E (2020) Effectiveness of passive design strategies in responding to future climate change for residential buildings in hot and humid Hong Kong. Energy Build 228:110469. https://doi.org/10.1016/J.ENBUILD.2020.110469
35. Vasaturo R, Van Hooff T, Kalkman I, Blocken B, Van Wesemael P (2018) Impact of passive climate adaptation measures and building orientation on the energy demand of a detached lightweight semi-portable building. In: Building simulation, vol 11, no 6. Tsinghua University Press, Beijing, pp 1163–1177. https://doi.org/10.1007/s12273-018-0470-8
36. Pajek L, Potočnik J, Košir M (2022) The effect of a warming climate on the relevance of passive design measures for heating and cooling of European single-family detached buildings. Energy Build 261:111947. https://doi.org/10.1016/J.ENBUILD.2022.111947
37. Durakovic B, Halilovic M (2023) Thermal performance analysis of PCM solar wall under variable natural conditions: an experimental study. Energy Sustain Dev 76:101274. https://doi.org/10.1016/J.ESD.2023.101274
38. Bamdad K (2023) Cool roofs: a climate change mitigation and adaptation strategy for residential buildings. Build Environ 236:110271. https://doi.org/10.1016/J.BUILDENV.2023.110271
39. Macintyre HL, Heaviside C (2019) Potential benefits of cool roofs in reducing heat-related mortality during heatwaves in a European city. Environ Int 127:430–441. https://doi.org/10.1016/J.ENVINT.2019.02.065
40. Hosseini M, Tardy F, Lee B (2018) Cooling and heating energy performance of a building with a variety of roof designs; the effects of future weather data in a cold climate. J Build Eng 17:107–114. https://doi.org/10.1016/j.jobe.2018.02.001
41. Virk G, Jansz A, Mavrogianni A, Mylona A, Stocker J, Davies M (2015) Microclimatic effects of green and cool roofs in London and their impacts on energy use for a typical office building. Energy Build 88:214–228. https://doi.org/10.1016/J.ENBUILD.2014.11.039

42. Andric I, Kamal A, Al-Ghamdi SG (2020) Efficiency of green roofs and green walls as climate change mitigation measures in extremely hot and dry climate: case study of Qatar. Energy Rep 6:2476–2489. https://doi.org/10.1016/J.EGYR.2020.09.006
43. Cirrincione L, Marvuglia A, Scaccianoce G (2021) Assessing the effectiveness of green roofs in enhancing the energy and indoor comfort resilience of urban buildings to climate change: methodology proposal and application. Build Environ 205:108198. https://doi.org/10.1016/J.BUILDENV.2021.108198
44. Yang Y, Javanroodi K, Nik VM (2021) Climate change and energy performance of European residential building stocks—a comprehensive impact assessment using climate big data from the coordinated regional climate downscaling experiment. Appl Energy 298:117246. https://doi.org/10.1016/J.APENERGY.2021.117246
45. Yan H, Ji G, Yan K (2022) Data-driven prediction and optimization of residential building performance in Singapore considering the impact of climate change. Build Environ 226:109735. https://doi.org/10.1016/J.BUILDENV.2022.109735
46. Huang KT, Hwang RL (2016) Future trends of residential building cooling energy and passive adaptation measures to counteract climate change: The case of Taiwan. Appl Energy 184:1230–1240. https://doi.org/10.1016/J.APENERGY.2015.11.008
47. Gupta R, Gregg M (2012) Using UK climate change projections to adapt existing English homes for a warming climate. Build Environ 55:20–42. https://doi.org/10.1016/J.BUILDENV.2012.01.014
48. Yu Z, Yang G, Zuo S, Jørgensen G, Koga M, Vejre H (2020) Critical review on the cooling effect of urban blue-green space: a threshold-size perspective. Urban Urban Green 49:126630. https://doi.org/10.1016/J.UFUG.2020.126630
49. Zuo J, Zhao ZY (2014) Green building research–current status and future agenda: a review. Renew Sustain Energy Rev 30:271–281. https://doi.org/10.1016/J.RSER.2013.10.021
50. Ascanio-Villabona J, Orejarena-Osorio N, Castillo-Leon NY, Tarazona-Romero BE, Jaimes-Quintero KT (2024) Thermal comfort evaluation using different glazing technologies in an energy model of rural housing in a cold climate. Herit Sustain Dev 6:735–748. https://doi.org/10.37868/HSD.V6I2.922
51. Moazami A, Carlucci S, Nik VM, Geving S (2019) Towards climate robust buildings: an innovative method for designing buildings with robust energy performance under climate change. Energy Build 202:109378. https://doi.org/10.1016/J.ENBUILD.2019.109378
52. Waddicor DA, Fuentes E, Sisó L, Salom J, Favre B, Jiménez C, Azar M (2016) Climate change and building ageing impact on building energy performance and mitigation measures application: a case study in Turin, Northern Italy. Build Environ 102:13–25. https://doi.org/10.1016/J.BUILDENV.2016.03.003
53. Rodrigues E, Fernandes MS (2020) Overheating risk in Mediterranean residential buildings: comparison of current and future climate scenarios. Appl Energy 259:114110. https://doi.org/10.1016/J.APENERGY.2019.114110
54. Jafarpur P, Berardi U (2021) Effects of climate changes on building energy demand and thermal comfort in Canadian office buildings adopting different temperature setpoints. J Build Eng 42:102725. https://doi.org/10.1016/J.JOBE.2021.102725
55. Domínguez-Amarillo S, Fernández-Agüera J, Sendra JJ, Roaf S (2019) The performance of Mediterranean low-income housing in scenarios involving climate change. Energy Build 202:109374. https://doi.org/10.1016/J.ENBUILD.2019.109374
56. Pérez-Andreu V, Aparicio-Fernández C, Martínez-Ibernón A, Vivancos JL (2018) Impact of climate change on heating and cooling energy demand in a residential building in a Mediterranean climate. Energy 165:63–74. https://doi.org/10.1016/J.ENERGY.2018.09.015
57. Liu S, Wang Y, Liu X, Yang L, Zhang Y, He J (2023) How does future climatic uncertainty affect multi-objective building energy retrofit decisions? Evidence from residential buildings in subtropical Hong Kong. Sustain Cities Soc 92:104482. https://doi.org/10.1016/J.SCS.2023.104482
58. Shen P, Braham W, Yi Y (2019) The feasibility and importance of considering climate change impacts in building retrofit analysis. Appl Energy 233–234:254–270. https://doi.org/10.1016/J.APENERGY.2018.10.041

59. D'Agostino D, Parker D, Epifani I, Crawley D, Lawrie L (2022) How will future climate impact the design and performance of nearly zero energy buildings (NZEBs)? Energy 240:122479. https://doi.org/10.1016/J.ENERGY.2021.122479
60. Zhang S, Ocłoń P, Klemeš JJ, Michorczyk P, Pielichowska K, Pielichowski K (2022) Renewable energy systems for building heating, cooling and electricity production with thermal energy storage. Renew Sustain Energy Rev 165. https://doi.org/10.1016/j.rser.2022.112560
61. Luo XJ, Oyedele LO (2022) Life cycle optimisation of building retrofitting considering climate change effects. Energy Build 258:111830. https://doi.org/10.1016/J.ENBUILD.2022.111830
62. Tarroja B, Chiang F, AghaKouchak A, Samuelsen S, Raghavan SV, Wei M, Sun K, Hong T (2018) Translating climate change and heating system electrification impacts on building energy use to future greenhouse gas emissions and electric grid capacity requirements in California. Appl Energy 225:522–534. https://doi.org/10.1016/J.APENERGY.2018.05.003
63. Berardi U, Jafarpur P (2020) Assessing the impact of climate change on building heating and cooling energy demand in Canada. Renew Sustain Energy Rev 121:109681. https://doi.org/10.1016/J.RSER.2019.109681
64. Santamouris M (2016) Cooling the buildings—past, present and future. Energy Build 128:617–638. https://doi.org/10.1016/J.ENBUILD.2016.07.034
65. Barea G, Victoria Mercado M, Filippín C, Monteoliva JM, Villalba A (2022) New paradigms in bioclimatic design toward climatic change in arid environments. Energy Build 266:112100. https://doi.org/10.1016/J.ENBUILD.2022.112100
66. Mutschler R, Rüdisüli M, Heer P, Eggimann S (2021) Benchmarking cooling and heating energy demands considering climate change, population growth and cooling device uptake. Appl Energy 288:116636. https://doi.org/10.1016/J.APENERGY.2021.116636
67. Li DHW, Yang L, Lam JC (2012) Impact of climate change on energy use in the built environment in different climate zones—a review. Energy 42:103–112. https://doi.org/10.1016/J.ENERGY.2012.03.044
68. Berger T, Amann C, Formayer H, Korjenic A, Pospichal B, Neururer C, Smutny R (2016) Impacts of external insulation and reduced internal heat loads upon energy demand of offices in the context of climate change in Vienna Austria. J Build Eng 5:86–95. https://doi.org/10.1016/J.JOBE.2015.11.005
69. Daly D, Cooper P, Ma Z (2014) Implications of global warming for commercial building retrofitting in Australian cities. Build Environ 74:86–95. https://doi.org/10.1016/J.BUILDENV.2014.01.008
70. Tian W, De Wilde P (2011) Uncertainty and sensitivity analysis of building performance using probabilistic climate projections: a UK case study. Autom Constr 20:1096–1109. https://doi.org/10.1016/J.AUTCON.2011.04.011
71. Wan KKW, Li DHW, Pan W, Lam JC (2012) Impact of climate change on building energy use in different climate zones and mitigation and adaptation implications. Appl Energy 97:274–282. https://doi.org/10.1016/J.APENERGY.2011.11.048
72. Duran-Sarmiento MA, Ascanio-Villabona JG, Tarazona-Romero BE (2024) Evaluation of model precision and performance discrepancies in simulated vs. experimental testing of vertical axis wind turbines. Sustain Eng Innov 6:411–420. https://doi.org/10.37868/SEI.V6I2.ID390.
73. Maduta C, D'Agostino D, Tsemekidi-Tzeiranaki S, Castellazzi L (2025) From nearly zero-energy buildings (NZEBs) to zero-emission buildings (ZEBs): current status and future perspectives. Energy Build 328:115133. https://doi.org/10.1016/J.ENBUILD.2024.115133
74. AlRubeei IRN, Idi SN, Hussein Alsammak IL, AlRikabi HT, Mutar HA, Alaidi AHM (2025) Using artificial intelligence for enhancement of solar cell efficiency in the of Iraq. Herit Sustain Dev 7:197–210. https://doi.org/10.37868/HSD.V7I1.1067
75. Ren Z, Paevere P, Chen D (2019) Feasibility of off-grid housing under current and future climates. Appl Energy 241:196–211. https://doi.org/10.1016/J.APENERGY.2019.03.068
76. Ghaffarianhoseini A, Dahlan ND, Berardi U, Ghaffarianhoseini A, Makaremi N, Ghaffarianhoseini M (2013) Sustainable energy performances of green buildings: a review of current theories, implementations and challenges. Renew Sustain Energy Rev 25:1–17. https://doi.org/10.1016/J.RSER.2013.01.010

77. Congedo PM, Baglivo C, D'Agostino D, Mazzeo D (2023) The impact of climate change on air source heat pumps. Energy Convers Manag 276:116554. https://doi.org/10.1016/J.ENCONMAN.2022.116554
78. Martins NR, Bourne-Webb PJ (2023) An assessment of the potential effect of climate change on renewable hybrid heating and cooling systems. J Build Eng 64:105598. https://doi.org/10.1016/J.JOBE.2022.105598
79. Jylhä K, Jokisalo J, Ruosteenoja K, Pilli-Sihvola K, Kalamees T, Seitola T, Mäkelä HM, Hyvönen R, Laapas M, Drebs A (2015) Energy demand for the heating and cooling of residential houses in Finland in a changing climate. Energy Build 99:104–116. https://doi.org/10.1016/J.ENBUILD.2015.04.001
80. Ohene E, Krarti M, Chan APC, Hsu SC, Ansah MK (2024) Optimal design guidelines for net zero energy residential buildings in cooling-dominated climates: case study of Ghana. Build Environ 260:111685. https://doi.org/10.1016/J.BUILDENV.2024.111685

Chapter 4
Energy Efficient Technologies for Zero Energy Buildings

Abstract This chapter overviews innovative envelope systems that improve energy efficiency in building, such as envelope insulation, application phase change materials, nano-enhanced composites and smart systems. Application of efficient lighting solutions and energy-optimized appliances can reduce peak loads and enhance occupant comfort. Overall, the chapter provides insights in implementing energy-efficient technologies that support the development of high-energy performance buildings as well as challenges associated with cost and availability of materials.

Keywords Energy-efficiency · Energy consumption · CO_2 emissions reduction · Energy loss reduction · Energy usage optimization

4.1 Introduction

The population growth and increased urbanization have created the significant need for energy use in buildings [1]. Recent studies show that buildings contribute nearly 40% of global energy consumption to heating, cooling, and lighting. As a result, many governments and organizations are implementing stricter regulations and standards to reduce energy demand and carbon footprint of new constructions. Furthermore, after the COVID-19 pandemic, which triggered a shift to remote work, a large portion of population began working from home, leading to an increase in energy consumption in buildings [2].

Energy efficiency serves as a cornerstone for advancing sustainability by addressing numerous certain challenges such as economic factors, minimizing waste and maximizing performance. In the context of Zero Energy Buildings (ZEBs), this means utilizing energy-efficient technologies and strategies to mitigate overall energy consumption [3], and the environmental impact [4]. Building envelope optimization, phase-change materials incorporations and improving efficiency of heating, ventilation, and air conditioning (HVAC) systems are some of the major measures [5]. Table 4.1 presents essential methods in terms of energy efficiency improvement in

B. Duraković, *Zero Energy Buildings*, Green Energy and Technology,
https://doi.org/10.1007/978-3-032-16027-0_4

buildings covering the use of energy-efficient materials, different smart appliances, renewable energy as well as designs for specific climate conditions.

Table 4.1 highlights the significance of advanced insulation materials and thermally efficient facades that reduce energy loss and CO_2 emissions. In addition, the table includes the integration of smart building technologies such as BMS that is required to optimize heating, cooling, lighting and improve the comfort for users. Occupant behavior, energy efficient materials, high-performance insulation, glazing, as well as efficiency of active systems contribute significantly to the overall building energy performance [4]. Energy efficiency can be optimized through different approaches including advanced materials [7], smart technologies and improved construction methods [11]. Vacuum-insulated panels and aerogel-based insulation offer exceptional thermal resistance, while electrochromic smart windows adjust their tint in response to sunlight, optimizing solar heat gain and cooling load reduction [17]. Emerging technologies, such as energy-generating windows and advanced insulation offer innovative solutions to optimize energy efficiency [8]. The integration of smart technologies and BMS [9] can enhance the overall energy performance by adjusting heating, cooling, and lighting based on real-time occupancy and environmental conditions.

Key challenges associated with their implementation are high initial costs [9], technical complexity, maintaining long-term efficiency [10], finding sustainable materials [6], and limited integration of renewable energy sources [12]. Table 4.2

Table 4.1 Summary of energy efficiency aspects in new buildings

Aspects	Results	Reference
Energy-efficient materials	• Advanced insulation materials and thermally efficient facades can reduce energy losses • Materials designed to reduce CO_2 emissions	[6, 7]
Smart technologies and systems	• Integration of smart building management systems (BMS) that optimizes heating, cooling, and lighting • Real-time adjustments improve energy use and comfort	[8–11]
Renewable energy	• Technologies like energy-generating windows reduce dependence on fossil fuels • It focuses on achieving net-zero emissions (by 2050)	[1, 2, 12, 13]
Design for hot and humid climates	• Energy-efficient designs adapted to climate conditions lowering energy consumption in buildings • Support through government and other initiatives in order to implement new technologies and practices	[14–16]

presents different challenges related to improving energy efficiency in building technologies. It covers issues encountered in areas like material selection, insulation, building envelop performance, windows and doors, and lighting systems.

Referring to Table 4.2 it is observed that the adoption and implementation of energy-efficient materials, insulation systems, advanced windows, and lighting technologies face certain barriers to being fully adopted. These barriers may include improper usage of materials, different financial and behavioral barriers of advanced technologies as well as their costs, and difficulties with implementing them.

Table 4.2 Different challenges related to improving energy efficiency in buildings

Category	Challenges	Reference
Materials and construction practices	• Improper selection of materials for temperature regulation • Leakage, oxidation and phase separation in PCMs • Limited market availability of effective switchable insulation products	[18–22]
Building envelope technologies	• Limited adoption of innovative materials like SIM and PCMs • Challenges in applying thermally switchable insulation due to operational differences • Energy losses through walls, roofs and windows	[23–30]
Reducing energy losses	• Significant heat loss through building envelopes like roofs, walls and foundations • Limited implementation of green roofs and advanced external wall technologies • High resource consumption and environmental impact of existing buildings like hotels	[24, 31–39]
Efficient windows and doors	• High initial costs of advanced glazing system • Integration of dynamic coatings with static glazing technologies • Irregular nature of renewable energy sources like solar	[40–47]
Lighting systems efficiency	• High upfront costs of adaptive lighting control systems • Limited adaption in developing regions due to resource constraints	[48–52]
Equipment efficiency	• Credit constraints and risk aversion in adopting efficient products • High costs of heat pump systems in cold regions • Lack of comprehensive programs to overcome behavioral and financial barriers	[53–60]

The purpose of this chapter is to analyze key methods and approaches based on existing research in the field of energy efficiency in ZEB and to investigate associated challenges [61]. The chapter focuses on how certain studies, through the implementation of energy-efficient materials such as special insulations, thermally efficient facades, and smart energy management systems, have successfully or attempted to reduce energy consumption in buildings and improved building energy efficiency [62]. Key aspects that are discussed here are energy-efficient materials and construction practices, insulation and building envelope technologies, high-performance windows and doors, energy-efficient lighting systems and appliances as well as strategies for reducing energy losses in the building envelope.

4.2 Materials and Construction Practices

Materials used in buildings are of great importance for achieving the building energy efficiency [7]. Material selection is a very important step, and improper choices can lead to significant inefficiencies. Ideally, materials should be chosen to help retain and regulate indoor temperature, thereby reducing reliance on artificial heating and cooling systems and minimizing overall energy consumption [18].

Recent high-performance materials like Vacuum Insulation Panels (VIPs) and Phase Change Materials (PCMs) actively regulate temperature by storing and releasing heat. This improves building thermal inertia and allows buildings to maintain more stable indoor conditions with less energy supply. For instance, application of VIPs in buildings significantly improves its thermal performance and enables thinner construction with comparable insulation levels [63].

VIPs are multi-layered panels that prevent the penetration of air and moisture, providing very high insulation efficiency and reducing heat transfer through walls. Recent experiments of next generation VIPs that incorporate nanoporous cores and multilayer metallized barrier films, indicate that this approach improved aging resistance, enabling durability in building application. One study experimented with incorporation of different core materials in VIPs, such as fumed silica, fiber-powder composites, and concluded that microstructure and compressive strength influence thermal performance and mechanical stability of the VIPs [64]. One study argued that, when exposed to different environments, barrier films also affect the durability of VIPs with fumed silica cores. This study suggests that polymeric film degrade faster, with a 20–60% increase in thermal conductivity over ~ 2 years under high humidity. On the contrary, metalized films only showed a 4–11% increase [65]. Improvement of VIPs leads to better energy performance and long-term economic benefits [21].

PCMs are also very effective in improving energy efficiency and reducing CO_2 emissions. When integrated into walls, roofs or floors, they can absorb and release heat as temperatures changes, helping to maintain stable indoor conditions and reducing the need for heating and cooling [20]. A study from 2018 shows that incorporating PCM into modular building walls can reduce the heat load by 10–30%, and at the same time significantly smooth room temperature fluctuations [66]. Another study examined the impact of PCM integrated into a traditional wall in Egypt on peak and average cooling energy consumption. The results indicate that a 2 cm PCM layer in a wall reduced interior surface temperature peaks by 2–5 °C and shifted the peak by 2–6 h, which helps align cooling demand with lower ambient temperatures [67].

A more advanced version, shape-stable phase change materials, can be obtained by combining PCMs with expanded graphite using a vacuum impregnation process. This improves both the thermal performance and stability of the materials, making them suitable for long-term applications in energy storage and management [68, 69].

A study from 2023 included examination of highly thermal conductive shape-stabilized composite phase change materials based on boron nitride and expanded graphite for solar thermal applications. The results indicate that a shape-stabilized PCM (stearic acid–benzamide eutectic) with 15 wt% BN + 20 wt% EG, achieved a thermal conductivity of ~ 6.99 W/m·K and excellent cycle stability (100 cycles) without significant leakage [70]. Another study explored combining BN nanosheets of different sizes with graphene oxide (GO), and concluded that this bond forms a thermal network that raises the PCM thermal conductivity to ~ 1.13 W/m·K, preserving good latent heat and shape stability at the same time [71]. Overall, advanced building materials such as VIPs and integration of PCMs into building envelopes enable an innovative approach to improving energy efficiency.

4.3 Innovations in Building Envelope Technologies

Reducing energy losses through building envelopes is one of the key strategies in energy efficiency improvement [31]. Modern approaches to building envelope design, which include dynamic systems capable of adapting to changes in the environment, demonstrated its potential for improving building energy performance [32].

Insulation of building envelopes is a fundamental measure in enhancing the control of the heat, moisture, and air flow between the interior and exterior [23], which helps to maintain stable indoor air comfort parameters [25]. This measure reduces the need for heating and cooling and consequently costs [24]. As insulation solutions, rigid foam insulation, laminated polymer glass and triple glazing are used. These materials significantly increase energy efficiency, especially where energy consumption exceeds 30% of global use [30].

The application of variable thermal insulation that has the ability to respond to changes in heating/cooling needs can improve the performance of conventional heating/cooling systems [27]. These technologies in practice are limited by the lack of commercially available products [27].

Building-integrated photovoltaic (BIPV) is a technology that is directly incorporated into building façades and serves at the same time as a thermal barrier and as a power generator. BIPV can be incorporated in roofs, façades, windows, shading devices, or skylights [72]. BIPV performs very good in warm and sunny climates, producing high energy yields, while BIPV windows in temperate regions help reduce heating and cooling loads. A major emerging trend is the shift toward flexible PV technologies, which are gaining attention over traditional silicon-based systems. Another review examines the electrical, optical, and thermal performance of PV windows and reports an average energy saving rate of about 37% for BIPV façades [73]. Challenges associated with BIPV that need to be addressed are still cost, integration complexity, limited standardization, and durability [72, 74].

A cutting-edge approach to improving thermal storage performance in building envelopes is the use of nano-enhanced phase change materials (NEPCMs). Incorporation of nanoparticles such as Al_2O_3, CuO, TiO_2, and carbon-based materials (e.g., graphene), reveal that thermal conductivity can be increased by 15–60%. The results also notice the decrease of latent heat 3–15% and propose careful selection of materials for proper optimization [75]. The synthesis methods, stability, and thermal performance under cyclic operation of NEPCMs built via emulsion polymerization maintain high encapsulation efficiency and good performance that that indicates a showing promising durability for real-world thermal energy storage [76]. Although very promising, the challenges related to nano-additives, such as nanoparticle agglomeration, reduced latent heat due to high particle loading, leakage, and cost, are still to be resolved.

The incorporation of PCMs into the building envelope, combined with night ventilation and thermal insulation layers, can significantly improve building thermal inertia [33, 37]. By properly selecting PCM materials, which operate in the temperature range of 18–32 °C, heat transfer through facades can be optimized [39, 77]. Biodegradable PCM materials based on renewable resources represent a more environmentally friendly alternative to traditional PCMs, but they still face challenges such as oxidation, phase separation and limited thermal conductivity which require further research [29].

Radiative cooling (RC) is an emerging technology in the field of passive strategies. Materials are emitting heat to the sky that acts like a cold sink, by reducing the temperature of the building surface without any added energy. These reflecting materials are suitable for the integration into roofs, facades, or windows, without compromising architectural and design aesthetics. The integration of radiative cooling with photovoltaics, thermal energy storage, and water harvesting offers additional features to building energy efficiency [78]. Challenges include material performance, scalability, and infrastructure integration [78].

Future research must provide enhancements and integration directions for BIPV, NEPCMs and similar innovative solutions to the building envelope design, to help reduce carbon emissions and energy consumption. Currently, the integration of BIPV comes with high initial costs, lack of standardized modules, and complex integration. Major barriers for NEPCMs remain scalable for manufacturing, long-term thermal cycling stability, and standard protocols for nanoparticle dispersion to avoid performance degradation. Future work must prioritize experimental and theoretical studies to address these challenges. While integration of radiative cooling into building applications shows high theoretical cooling power, real-world implementation needs to carefully consider geographic and climatic constraints and long-term degradation.

4.4 Efficient Windows and Doors

High-performance windows and doors can greatly reduce heating and cooling demands, which can decrease energy consumption as well as greenhouse gas emissions [40]. These technologies improve energy efficiency and comfort in buildings through its features such as multi-layer glazing [79], low-emissivity (low-e) coatings [41], insulated frames [42], and gas-filled spaces between the glass panes [43], PCM based glazing [80], which tend to minimize heat transfer or reduce indoor temperature fluctuation. In addition, these products give benefits like noise reduction, security, and good natural light, contributing to the overall well-being of building occupants [44].

The importance of natural ventilation is essential to mention because of its impact on indoor air parameters as well as occupant thermal comfort [45]. Dye-sensitized solar cells (DSSCs) are promising technology for incorporation within windows due to their potential for low-cost fabrication and versatile applications, including aesthetic appearance, semi-transparency, and color tunability [46]. When the power generation efficiency of DSSC exceeds 13%, its overall energy consumption becomes lower than that of the Low-e windows. Energy performance of a two-story building in Morocco's semi-arid climate focusing on the impact of various static and dynamic glazing types on energy consumption, thermal and visual comfort, and environmental and economic factors is evaluated. Static double glazing with low-e coatings is the most energy-efficient option, while integrating thermochromics and electrochromic dynamic glazing technologies [47].

Another study explores the impact of integrating dynamic coatings, such as thermochromics and electrochromic films, with static glazing on building energy efficiency and indoor comfort in a semi-arid climate. Using a numerical model with Energy Plus and Radiance tools, various window configurations were evaluated for their effects on energy consumption, thermal and visual comfort, and environmental impact. The results show that double glazing, combined with dynamic coatings, significantly improves energy efficiency, reducing energy consumption by 26% and emissions by 1522.38 kgCO_2/year, with a payback period of 12.86 years.

The integration of partially filled PCM triple glazed windows shows that it effectively regulates indoor temperatures by storing and releasing thermal energy, improving occupant comfort, and reducing building energy demand. Field tests in Korea showed daytime temperature reductions of ~ 2 °C and nighttime increases of ~ 1.5 °C. In this case, seasonal simulations achieved up to 3.7 °C cooling during the day and 2 °C warming at night [81]. Small-scale experiments in this study gave promising results; however, optimal PCM melting temperature, installation capacity, and real-world factors like ventilation and internal loads need to be considered for full-scale building applications. A study from Chile presented that one-year real-scale experiment in two east-facing office buildings showed PCM-integrated glazing increases thermal inertia, reducing HVAC energy consumption and peak cooling loads during summer and mid-seasons. It also improved indoor thermal comfort by stabilizing mean radiant temperatures year-round and enhanced visual comfort by improving luminance distribution in winter and cooler months [82]. Innovative sliding smart windows are also being experimented with. The results show that integrating air-gap, phase change material (PCM), photovoltaic (PV), and vacuum glazing (VG) technologies to simultaneously provide electricity generation, thermal insulation, and heat storage can significantly reduce solar heat gains. This study highlighted that this design lead to generating up to ~ 1.4 kWh of PV electricity per square meter per day, highlighting the effectiveness of combining PCM, PV, and VG for zero-energy building applications [83] A study from 2022 reviewed hybrid BIPV and Building-Integrated Concentrating Photovoltaic (BICPV) glazing systems, including smart glazing technologies that incorporate thermotropic hydrogels and hydroxypropyl cellulose for dynamic daylight and solar heat control. Some of the challenges related to this innovative approach include optimizing dynamic control under variable weather conditions, balancing transparency vs. light scattering, ensuring long-term material stability, and integrating these systems effectively into building facades [84].

4.5 Lighting Systems Efficiency

Energy-efficient lighting systems play a key role in reducing electricity energy consumption and minimizing environmental impact [48]. Therefore, lighting systems became an essential strategy in residential and commercial building energy demand reductions optimizing the use of light and lowering electricity [49, 85]. A study from 2024 examined the effect of indoor lighting energy reduction using a daylight-concentrating indoor louver system. They found that the daylight-concentrating indoor louvers combined with LED dimming control can dramatically reduce lighting energy use, achieving up to 85.65% savings in transitional seasons, due to better natural light penetration. However, the winter performance was weaker because of lower daylight availability, resulting in about 25.85% higher power consumption. Overall, the system improves indoor illuminance distribution and significantly reduces electric lighting demand, especially when solar altitude supports deeper daylighting [86].

A smart lighting control system that is designed to reduce energy consumption in office buildings automatically adjusts lighting based on changes in daylight, occupancy, and user preferences. With the sensors at each workstation, this system adjusts the relationship between dimming levels and illuminance. The adaptive control algorithm ensures consistent lighting while it minimizes energy usage. These results show that the system reduces energy consumption by approximately 40% in comparison to conventional LED lighting systems [51]. Advanced lighting proved to light energy savings to 77%, by a daylight-responsive dimming system (DRDS) improved with a new control algorithm that accounts for indirect illuminance. The full-scale test, held for over 23 days, further solved accuracy problems often present in methods of traditional calibration, and achieved 98.9% illuminance accuracy. This proved the effectiveness of innovative methods for sustainable building lighting [87].

An energy-efficient lighting system in office buildings combines natural daylight harvesting with occupancy sensors to dynamically adjust artificial lighting. This intelligent control strategy optimizes lighting levels based on natural light availability, room occupancy, and significantly reduces overall energy consumption [88].

A mixed strategy that is a combination of occupancy-based and illuminance-based approaches may serve as an optimal solution within commercial building lighting systems. This method utilizes sensors to detect occupancy and adjust lighting levels based on natural daylight while keeping visual comfort [52]. Recent studies on occupancy-based lighting control in offices examined the energy saving potential of lighting based on the occupancy patterns. The study was comparing room-level versus desk-level control, and the results indicate that while variations in individual occupancy patterns have minimal effect on relative energy savings, desk-level control consistently provides higher energy savings than coarser strategies. The implementation of fine-grained occupancy sensors within mixed strategies supports more flexible and responsive lighting control in commercial environments [89].

4.6 Appliances and Equipment Efficiency

Energy-efficient appliances and equipment are used to achieve less energy consumption while providing the same performance compared to conventional alternatives [53]. They are designed with advanced features such as smart controls, better insulation, as well as optimized operational efficiency that ensures reduced energy waste [58]. The appliances are crucial for reduction of energy consumption by lowering utility costs [54] and minimizing possible environmental impact [55, 90].

One article highlights that despite California's goal to double energy efficiency by 2030, traditional approaches like codes and utility programs have failed to encourages widespread adoption of efficient technologies. They identified five barriers that prevent customers from adopting energy-efficient products: credit constraints, risk aversion, imperfect information, myopic behavior, and externalities. In this article, there is a case study of Puget Sound Energy (PSE), which proposes a leasing program to promote energy-efficient appliances for residential and commercial customers. This program includes heat pumps and water heaters, with lifetime maintenance, emphasizing the potential public benefits of increased adoption of energy-efficient products. The PSE survey showed a strong customer's desire to lease energy-efficient appliances. By giving financial flexibility and mitigating risks, utility-led leasing programs can improve the adoption of energy-efficient products, addressing gaps left by traditional energy efficiency programs [59].

One of the possible solutions to the above problem is to replace fossil fuel heating systems with heat pump systems. Air-source heat pumps with vapor compression technology are simple and have low costs but have a disadvantage in cold regions. In cold regions, three types of vapor compression systems are commonly used: single-stage, dual-stage, and multi-stage. Single-stage systems use advanced refrigerants and ejectors to improve efficiency; dual-stage systems divide the compression process for better performance; and multi-stage systems enhance the coefficient of performance using injection ports [60]. To maximize the benefits of these advanced heat pump systems, integrating them with smart control technologies such as BMS can further optimize energy use and improve overall building performance [91].

4.7 Conclusions

The incorporation of green building practices and energy-efficient appliances supports efforts toward sustainability goals. PCMs, vacuum-insulated panels, smart lighting systems, and dynamic glazing are examples of technologies that can improve building energy performance and thermal comfort. Challenges, such as technical complexity, costs, limited market availability, and long-term material performance, must be addressed in the future. Therefore, collaborative work across sectors is required to find optimum solutions.

References

1. Naylor S, Gillott M, Lau T (2018) A review of occupant-centric building control strategies to reduce building energy use. Renew Sustain Energy Rev 96:1–10. https://doi.org/10.1016/J.RSER.2018.07.019
2. Rana A, Kamali M, Riyadh MM, Sultana SR, Kamal MR, Alam MS, Hewage K, Sadiq R (2022) Energy efficiency in residential buildings amid COVID-19: a holistic comparative analysis between old and new normal occupancies. Energy Build 277:112551. https://doi.org/10.1016/J.ENBUILD.2022.112551
3. Jung W, Jazizadeh F (2019) Human-in-the-loop HVAC operations: a quantitative review on occupancy, comfort, and energy-efficiency dimensions. https://doi.org/10.1016/j.apenergy.2019.01.070
4. Wang Y, Hu B, Meng X, Xiao R (2024) A Comprehensive review on technologies for achieving zero-energy buildings. Sustainability 16:10941. https://doi.org/10.3390/SU162410941
5. Pérez-Lombard L, Ortiz J, Maestre IR, Coronel JF (2012) Constructing HVAC energy efficiency indicators. Energy Build 47:619–629. https://doi.org/10.1016/j.enbuild.2011.12.039
6. Firoozi AA, Firoozi AA, Oyejobi DO, Avudaiappan S, Flores ES (2024) Emerging trends in sustainable building materials: technological innovations, enhanced performance, and future directions. Results Eng 24:103521. https://doi.org/10.1016/J.RINENG.2024.103521
7. Shao L (2010) Materials for energy efficiency and thermal comfort in new buildings. Mater Energy Effic Therm Comf Build 631–648. https://doi.org/10.1533/9781845699277.3.631
8. Esmaeilzadeh A, Zakerzadeh MR, Koma AY (2018) The comparison of some advanced control methods for energy optimization and comfort management in buildings. Sustain Cities Soc 43:601–623. https://doi.org/10.1016/J.SCS.2018.08.038
9. Bilous I, Biriukov D, Karpenko D, Eutukhova T, Novoseltsev O, Voloshchuk V (2024) Reinforcement learning model for energy system management to ensure energy efficiency and comfort in buildings. Energy Eng 121:3617–3634. https://doi.org/10.32604/EE.2024.051684.
10. Pajek L, Košir M (2021) Strategy for achieving long-term energy efficiency of European single-family buildings through passive climate adaptation. Appl Energy 297:117116. https://doi.org/10.1016/J.APENERGY.2021.117116

11. Zhang G, Wu H, Liu J, Liu Y, Ding Y, Huang H (2024) A review on switchable building envelopes for low-energy buildings. Renew Sustain Energy Rev 202:114716. https://doi.org/10.1016/J.RSER.2024.114716
12. Abed AM, Alhuyi Nazari M, Ahmadi MH, Mukhtar A, Kumar R, Gharib N (2025) Power generation by utilization of different renewable energy sources in five Middle Eastern countries: present status, opportunities and challenges. Sustain Energy Technol Assess 73:104101. https://doi.org/10.1016/J.SETA.2024.104101
13. Khalifa F, Marzouk M (2025) Integrated blockchain and digital twin framework for sustainable building energy management. J Ind Inf Integr 43:100747. https://doi.org/10.1016/J.JII.2024.100747
14. Mohammad Yusoff WF, Mohamed MF (2024) Building energy efficiency in hot and humid climate. Encycl Sustain Technol 214–229. https://doi.org/10.1016/B978-0-323-90386-8.00048-6
15. Byers BS, Raghu D, Olumo A, De Wolf C, Haas C (2024) From research to practice: a review on technologies for addressing the information gap for building material reuse in circular construction. Sustain Prod Consum 45:177–191. https://doi.org/10.1016/J.SPC.2023.12.017
16. Nord N (2024) Building energy efficiency in cold climates. encyclopedia of sustainable technologies, pp 205–213. https://doi.org/10.1016/B978-0-323-90386-8.00047-4
17. What innovative technologies are revolutionizing the creation of net zero buildings? https://zerocarbonmission.com/innovative-technologies-behind-net-zero-buildings/?utm_source=chatgpt.com. Last accessed 22 May 2025
18. Lu Y, Yang Z, Yu J, Chen B, Zhong K (2021) Development of a second-order dynamic model for quantifying impact of thermal mass on indoor thermal environment. J Build Eng 42:102496. https://doi.org/10.1016/J.JOBE.2021.102496
19. Saleem AM, Badr AA, Alyas BH, Alomar OR (2024) Performance of thermal insulation of different composite walls and roofs materials used for energy efficient building construction in Iraq. Front Heat Mass Transf 22:1231–1244. https://doi.org/10.32604/FHMT.2024.053770
20. Zhao S, Ming T, Wu Y, Cai C, Yin K, Fang Y, de Richter R, Chen Y, Zhou N (2024) Application research and effectiveness analysis of phase change materials in building envelope: a review. Energy Build 324:114923. https://doi.org/10.1016/J.ENBUILD.2024.114923
21. Ashraf Ali N, Sadek S, Abdin A (2024) A review on innovative nanomaterials for enhancing energy performance of the building envelope. Curr Nanomater 9:287–302. https://doi.org/10.2174/0124054615248038231020054831
22. Sun Q, Du C, Chen G (2025) Thermoelectric materials and devices: applications in enhancing building energy conversion and efficiency. Adv Nanocomposites 2:15–31. https://doi.org/10.1016/J.ADNA.2024.11.001
23. Li H, Liu M, Cui Q, Kong X, Jia J, Fan M (2024) Optimization research on the air infection intervention ability of interactive cascade ventilation in high-density building spaces. J Build Eng 98:111376. https://doi.org/10.1016/J.JOBE.2024.111376
24. Cao L, Li T, Wang F (2024) Heat insulation and thermal insulation method of passive low energy consumption residential building exterior envelope structure based on BIM. Results Eng 23:102734. https://doi.org/10.1016/J.RINENG.2024.102734
25. Wang KJ, Lin CJ, Dagne TB (2023) An adaptive indoor temperature control approach simultaneously improving thermal comfort and task performance. Int J Therm Sci 193:108542. https://doi.org/10.1016/J.IJTHERMALSCI.2023.108542
26. Olabi AG, Shehata N, Issa UH, Mohamed OA, Mahmoud M, Abdelkareem MA, Abdelzaher MA (2025) The role of green buildings in achieving the sustainable development goals. Int J Thermofluids 25:101002. https://doi.org/10.1016/J.IJFT.2024.101002
27. Cui H, Overend M (2019) A review of heat transfer characteristics of switchable insulation technologies for thermally adaptive building envelopes. https://doi.org/10.1016/j.enbuild.2019.07.004
28. Santamouris M, Vasilakopoulou K Present and future energy consumption of buildings: challenges and opportunities towards decarbonisation. e-Prime—Adv Electr Eng Electron Energy. 1:100002 (2021). https://doi.org/10.1016/J.PRIME.2021.100002

29. Malode SJ, Shetti NP (2025) Thermal energy storage systems using bio-based phase change materials: a comprehensive review for building energy efficiency. J Energy Storage 105:114709. https://doi.org/10.1016/J.EST.2024.114709
30. Kapoor G, Singhal M (2024) Impact of innovative thermal insulation materials in the building envelope on energy efficiency of residential buildings. Mater Today Proc. https://doi.org/10.1016/J.MATPR.2024.04.041
31. Ferrara M, Fabrizio E, Virgone J, Filippi M (2016) Energy systems in cost-optimized design of nearly zero-energy buildings. Autom Constr 70:109–127. https://doi.org/10.1016/J.AUTCON.2016.06.007
32. Zhao T, Zhuang Z, Chang T, Grunewald J, Yu Y (2025) Performance-oriented design of thermal-activated building envelope based on multi-objective optimization: Method and application. Energy Build 328:115198. https://doi.org/10.1016/J.ENBUILD.2024.115198
33. Al-Yasiri Q, Alktranee M, Szabó M, Arıcı M (2023) Building envelope-enhanced phase change material and night ventilation: effect of window orientation and window-to-wall ratio on indoor temperature. Renew Energy 218:119263. https://doi.org/10.1016/J.RENENE.2023.119263
34. Lozinsky CH, Touchie MF (2023) Comparing air sealing techniques for suite compartmentalization in a multi-unit residential building: aerosolized sealant vs. conventional air sealing approaches. J Build Eng 71:106528. https://doi.org/10.1016/J.JOBE.2023.106528
35. El Hassnaoui A, Sakami S, Abddaim E, Boukhattem L, Ait Nouh F, Benhamou B (2024) Experimental evaluation of building energy efficiency through innovative eco-friendly roofing materials in semi-arid climates. Energy Build 309:114054. https://doi.org/10.1016/J.ENBUILD.2024.114054
36. Wang Y, Wang ZH, Rahmatollahi N, Hou H (2024) The impact of roof systems on cooling and building energy efficiency. Appl Energy 376:124339. https://doi.org/10.1016/J.APENERGY.2024.124339
37. Asghari M, Fereidoni S, Fereidooni L, Nabisi M, Kasaeian A (2024) Energy efficiency analysis of applying phase change materials and thermal insulation layers in a building. Energy Build 312:114211. https://doi.org/10.1016/J.ENBUILD.2024.114211
38. Sozer H (2010) Improving energy efficiency through the design of the building envelope. Build Environ 45:2581–2593. https://doi.org/10.1016/J.BUILDENV.2010.05.004
39. Azadani HN, Ardekani A (2025) Effect of applying phase change materials (PCM) in building facades on reducing energy consumption. Next Energy 6:100210. https://doi.org/10.1016/J.NXENER.2024.100210
40. Imeri A, Gil-Alana LA (2024) Persistence in greenhouse gas emissions: evidence from European countries. Energy Rep 12:5793–5800. https://doi.org/10.1016/J.EGYR.2024.11.049
41. Wurm J, Fujisawa-Phillips ST, Rasskazov IL (2025) Optimal design of low-emissivity coatings. Sol Energy Mater Sol Cells 280:113267. https://doi.org/10.1016/J.SOLMAT.2024.113267
42. Hietikko J, Tuominen E, Valovirta I, Vinha J (2024) Timber-framed exterior walls insulated with wood shavings: a field study in a Nordic climate. Build Environ 254:111371. https://doi.org/10.1016/J.BUILDENV.2024.111371
43. Jelle BP (2011) Traditional, state-of-the-art and future thermal building insulation materials and solutions—properties, requirements and possibilities. Energy Build 43:2549–2563. https://doi.org/10.1016/J.ENBUILD.2011.05.015
44. Ono E, Chong A, Lam KP (2025) Occupant-centric zoning design framework for flexible workplaces: Impact of occupancy modeling and building operation. Build Environ 267:112294. https://doi.org/10.1016/J.BUILDENV.2024.112294
45. Kocik S, Psikuta A, Ferdyn-Grygierek J (2024) Influence of window and door opening on office room environment and human thermal sensation during different seasons in moderate climate. Build Environ 259:111669. https://doi.org/10.1016/J.BUILDENV.2024.111669
46. Hyun JY, Park BR, Kim NH, Moon JW (2022) Building energy performance of DSSC BIPV windows in accordance with the lighting control methods and climate zones. Sol Energy 244:279–288. https://doi.org/10.1016/J.SOLENER.2022.08.039
47. Es-sakali N, Idrissi Kaitouni S, Ait Laasri I, Oualid Mghazli M, Cherkaoui M, Pfafferott J (2024) Static and dynamic glazing integration for enhanced building efficiency and indoor

comfort with thermochromic and electrochromic windows. Therm Sci Eng Prog 52:102681. https://doi.org/10.1016/J.TSEP.2024.102681
48. Uyan E, Atlar M, Ölçer AI (2023) Improving the energy efficiency of lighting systems for a marine equipment manufacturing plant through retrofitting, daylighting, and behaviour change. J Clean Prod 413:137216. https://doi.org/10.1016/J.JCLEPRO.2023.137216
49. Upadhyay P, Kuchhal P, Mondal S (2024) A review of the use of different technologies/methods for the transmission of solar radiation for lighting purposes using optical fibers. Renew Energy Focus 50:100614. https://doi.org/10.1016/J.REF.2024.100614
50. Chen TC, Dwijendra NKA, Singhal S, Sivaraman R, Mamdouh A (2023) Intelligent system application to monitor the smart city building lighting. Comput Mater Contin 75:3159–3169. https://doi.org/10.32604/CMC.2023.035418
51. Aussat Y, Rosmanis A, Keshav S (2022) A power-efficient self-calibrating smart lighting system. Energy Build 259:111874. https://doi.org/10.1016/J.ENBUILD.2022.111874
52. Wagiman KR, Abdullah MN, Hassan MY, Mohammad Radzi NH, Abu Bakar AH, Kwang TC (2020) Lighting system control techniques in commercial buildings: current trends and future directions. J Build Eng 31:101342. https://doi.org/10.1016/J.JOBE.2020.101342
53. Javaid S, Fahim H, Zeadally S, He B (2024) From sensing to energy savings: a comprehensive survey on integrating emerging technologies for energy efficiency in WBANs. Digit Commun Netw. https://doi.org/10.1016/J.DCAN.2024.11.012
54. Liu Y, Guo X, Hu F (2014) Cost-benefit analysis on green building energy efficiency technology application: a case in China. Energy Build 82:37–46. https://doi.org/10.1016/J.ENBUILD.2014.07.008
55. Kavousian A, Rajagopal R, Fischer M (2015) Ranking appliance energy efficiency in households: Utilizing smart meter data and energy efficiency frontiers to estimate and identify the determinants of appliance energy efficiency in residential buildings. Energy Build 99:220–230. https://doi.org/10.1016/J.ENBUILD.2015.03.052
56. Duan Z, de Wilde P, Attia S, Zuo J (2024) Prospect of energy conservation measures (ECMs) in buildings subject to climate change: a systematic review. Energy Build 322:114739. https://doi.org/10.1016/J.ENBUILD.2024.114739
57. Paramati SR, Shahzad U, Doğan B (2022) The role of environmental technology for energy demand and energy efficiency: evidence from OECD countries. Renew Sustain Energy Rev 153:111735. https://doi.org/10.1016/J.RSER.2021.111735
58. Coyne B, Denny E, Fitiwi DZ (2023) The benefits of low-carbon energy efficiency technology adoption for data centres. Energy Convers Manag: X 20:100447. https://doi.org/10.1016/J.ECMX.2023.100447
59. Faruqui A, Lessem N, Trewn H (2017) The public benefits of leasing energy efficient equipment. Electr J 30:8–16. https://doi.org/10.1016/J.TEJ.2017.06.005
60. Zhang L, Jiang Y, Dong J, Yao Y (2018) Advances in vapor compression air source heat pump system in cold regions: a review. Renew Sustain Energy Rev 81:353–365. https://doi.org/10.1016/J.RSER.2017.08.009
61. Himeur Y, Alsalemi A, Al-Kababji A, Bensaali F, Amira A, Sardianos C, Dimitrakopoulos G, Varlamis I (2021) A survey of recommender systems for energy efficiency in buildings: principles, challenges and prospects. Inf Fusion 72:1–21. https://doi.org/10.1016/J.INFFUS.2021.02.002
62. Yang Y, Chen F (2023) Research on energy-saving coupling correlation of new energy buildings based on carbon emission effect. Sustain Energy Technol Assess 56:103043. https://doi.org/10.1016/J.SETA.2023.103043
63. Baetens R, Jelle BP, Thue JV, Tenpierik MJ, Grynning S, Uvsløkk S, Gustavsen A (2010) Vacuum insulation panels for building applications: a review and beyond. Energy Build 42:147–172. https://doi.org/10.1016/J.ENBUILD.2009.09.005
64. Mukhopadhyaya P, Kumaran MK, Normandin N, van Reenen D, Lackey JC (2008) High-performance vacuum insulation panel: development of alternative core materials. J Cold Regions Eng 22

65. Aldykiewicz AJ Jr, Desjarlais AO, Biswas K (2022) The effect of barrier films and exposure on the aging of vacuum insulation panels with fumed silica cores. In: IVIS 2021 15th International vacuum insulation symposium—11–12 April 2022
66. Gobinath S, Senthilkumar G, Beemkumar N (2018) Comparative study of room temperature control in buildings with and without the use of PCM in walls. Energy Sources Part A: Recover, Util Environ Eff 40:1765–1771. https://doi.org/10.1080/15567036.2018.1486910;PAGE:STRING:ARTICLE/CHAPTER
67. Kamel JA, Mina EM, Elsabbagh AM (2022) Implementation of phase change material for cooling load reduction: a case study for Cairo, Egypt. Int J Air-Cond Refrig 30(1):13. https://doi.org/10.1007/S44189-022-00013-5
68. Yadav A, Samykano M, Pandey AK, Rajamony RK, Tyagi VV (2024) Thermal characterization of shape-stable phase change material for efficient thermal energy storage and electric to thermal energy conversion. J Energy Storage. 103:114368. https://doi.org/10.1016/J.EST.2024.114368
69. Duraković B (2024) Phase change material applications in construction and building materials. In: Phase change materials for thermal energy management and storage, pp 231–260. https://doi.org/10.1201/9781003331957-10.
70. Huang D, Ma G, Yu Z, Lv P, Zhou Q, Liu Q, Peng C, Xiong F, Huang Y (2023) Highly thermal conductive shape-stabilized composite phase change materials based on boron nitride and expanded graphite for solar thermal applications. RSC Adv 13:13252–13262. https://doi.org/10.1039/D3RA01166F
71. Xin S, Zhao Z, Liu S, Liu J, Li M (2023) Study on the enhancement effect of synergy between multi-size functionalized boron nitride and graphene oxide on the thermal properties of phase change composites. Molecules 28:3797. https://doi.org/10.3390/MOLECULES28093797
72. Zarate-Perez E, Grados J, Rubiños S, Grados-Espinoza H, Astocondor-Villar J, Zarate-Perez E, Grados J, Rubiños S, Grados-Espinoza H, Astocondor-Villar J (2023) Building-integrated photovoltaic (BIPV) systems: a science mapping approach. AIMS Energy 11(6):1131–1152. https://doi.org/10.3934/ENERGY.2023052
73. Mohamed A, Ventura C, Shi S, Zhu N (2023) Challenges and optimization of building-integrated photovoltaics (BIPV) windows: a review. Sustainability 15:15876. https://doi.org/10.3390/SU152215876
74. Fragulis GF, Ilic P, Narputro P, Artiyasa M, Saputri US, Damas Permadi D, Hidayat M, Sari NK, Marifah SL (2025) Building-integrated photovoltaics: a bibliometric review of key developments and knowledge gaps. Eng Proc 107:36. https://doi.org/10.3390/ENGPROC2025107036
75. Narbuts J, Vanaga R (2025) Nano-enhanced phase change materials for building envelopes: a systematic review of thermal performance and applications. Environ Clim Technol. 29:359–389. https://doi.org/10.2478/RTUECT-2025-0025
76. Khlissa F, Mhadhbi M, Aich W, Hussein AK, Alhadri M, Selimefendigil F, Öztop HF, Kolsi L (2023) Recent advances in nanoencapsulated and nano-enhanced phase-change materials for thermal energy storage: a review. Processes 11:3219. https://doi.org/10.3390/PR11113219
77. Duraković B (2020) Phase change materials for building envelope. In: PCM-based building envelope systems: innovative energy solutions for passive design. Springer Nature, pp 17–37. https://doi.org/10.1007/978-3-030-38335-0_2
78. Chen H, Liu X, Liu J, Wang F, Wang C (2025) Radiative cooling applications toward enhanced energy efficiency: system designs, achievements, and perspectives. The Innovation 6:100999. https://doi.org/10.1016/J.XINN.2025.100999
79. Pascual C, Overend M (2021) Buckling of sandwich struts with a particular application in composite multi-layer glazing. Eng Struct 230:111545. https://doi.org/10.1016/J.ENGSTRUCT.2020.111545
80. Durakovic B, Torlak M (2017) Experimental and numerical study of a PCM window model as a thermal energy storage unit. Int J Low-Carbon Technol 12:272–280. https://doi.org/10.1093/ijlct/ctw024
81. Song K, Lee R, Kim D, Yoon J, Shin D (2025) Thermal performance evaluation of phase change material-integrated triple-glazed windows under Korean climatic conditions. Energies (Basel) 18. https://doi.org/10.3390/EN18215754

82. Uribe D, Vera S (2021) Assessment of the effect of phase change material (PCM) glazing on the energy consumption and indoor comfort of an office in a semiarid climate. Appl Sci 11:9597. https://doi.org/10.3390/APP11209597
83. Ahmed M, Radwan A, Serageldin A, Memon S, Katsura T, Nagano K (2020) Thermal analysis of a new sliding smart window integrated with vacuum insulation, photovoltaic, and phase change material. Sustainability 12:7846. https://doi.org/10.3390/SU12197846
84. Liu X, Wu Y (2022) A review of advanced architectural glazing technologies for solar energy conversion and intelligent daylighting control. Arch Intell 2022 1(1):10. https://doi.org/10.1007/S44223-022-00009-6
85. Alketbi MAS, Mushtaha ES (2025) The impact of artificial lighting on improving the visual experience of museum visitors: a case study of Sharjah Museum of Islamic Civilization. Herit Sustain Dev 7:919–940. https://doi.org/10.37868/HSD.V7I2.1271
86. Lee JH, Kang JS (2024) Study on lighting energy savings by applying a daylight-concentrating indoor louver system with LED dimming control. Energies 17:3425. https://doi.org/10.3390/EN17143425
87. Kim IT, Kim YS, Nam H, Hwang T (2018) Advanced dimming control algorithm for sustainable buildings by daylight responsive dimming system. Sustainability 10:4087. https://doi.org/10.3390/SU10114087
88. Soori PK, Vishwas M (2013) Lighting control strategy for energy efficient office lighting system design. Energy Build 66:329–337. https://doi.org/10.1016/J.ENBUILD.2013.07.039
89. De Bakker C, Van De Voort T, Rosemann A (2017) The energy saving potential of occupancy-based lighting control strategies in open-plan offices: the influence of occupancy patterns. Energies 11:2. https://doi.org/10.3390/EN11010002
90. Tarazona-Romero BE, Socha-Rojas ND, Ascanio-Villabona J, Kafarov V, Castillo-Leon NY (2024) Performance evaluation of solar thermal collectors in Colombian thermal floors by dynamic simulation. Sustain Eng Innov 6:421–432. https://doi.org/10.37868/SEI.V6I2.ID394
91. Liang R, Wang PH (2024) Enhancing energy efficiency in buildings, optimization method and building management systems application for lower CO_2 emissions. Energy 313:134054. https://doi.org/10.1016/J.ENERGY.2024.134054

Chapter 5
Energy Storage Technologies

Abstract Increasing renewable energy participation in total energy supply has faced its variability challenges that affect grid variability, reliability and efficiency. Energy storage solutions enable the surplus energy to be captured, converted and reused as needed, by reducing demand variability. This chapter provides a summary of technologies used in building energy storage, including their primary types, techno-economic considerations, and environmental impact. It investigates the possible advantages and disadvantages of incorporating energy storage devices.

Keywords Energy storage technologies · Techno-economic analysis · Environmental impact · Social impact · Power grid integration · Future trends

5.1 Introduction

A major source of electricity generation today is fossil fuels. This results in depletion of these sources and the emission of harmful pollutants and greenhouse gases. Renewable energy sources such as solar and wind power have experienced notable expansion and participation in energy supply. Due to their changing nature, electrical grid faces challenges that affect its reliability, resilience, and operational efficiency. However, their practical application appears promising in the coming years due to anticipated cost reductions and rapid advancements in emerging energy storage technologies [1].

The process of converting and storing electricity from a power grid that can be recovered and used when needed is known as electrical energy storage. This procedure makes it possible to generate electricity from variable sources of energy, during reduced usage times, or at low production costs. At instances of high energy demand and high costs, or when other generation options are unavailable, the stored energy can then be use [2].

The objective of this chapter is to review the current energy storage solutions such as batteries, pumped hydro, thermal storage, thermochemical heat storage, flywheel

B. Duraković, *Zero Energy Buildings*, Green Energy and Technology,
https://doi.org/10.1007/978-3-032-16027-0_5

storage. Some of them are currently available and some are still under development. There will be a discussion of performance, evaluation, comparison of existing technologies, challenges, and opportunities, as well as potential development and research of these technologies.

5.2 Current Energy Storage Solutions

Distributed generation technologies are becoming more popular as needs for clean energy grow. While these technologies can help address global warming, their less predictable and adjustable power output presents challenges in meeting changing demand cycles compared to traditional sources. To overcome this limitation, the most practical solution is to store the excess power generated by distributed production using sustainable energy sources.

To ensure that distributed generation renewable energy sources serve as a reliable primary energy supply, energy storage becomes an essential component of this system [3]. Some of the examples of energy storage is a wind-up clock, which serves as a mechanical energy storage device. In computers, batteries store chemical energy that is readily convertible, ensuring the clock chip remains operational even after turning off the computer. Another example is a hydroelectric storage system, which stores potential gravitational energy as power in a reservoir [4]. Figure 5.1 represents some of the common energy storage forms for used in Zero Energy Buildings (ZEB).

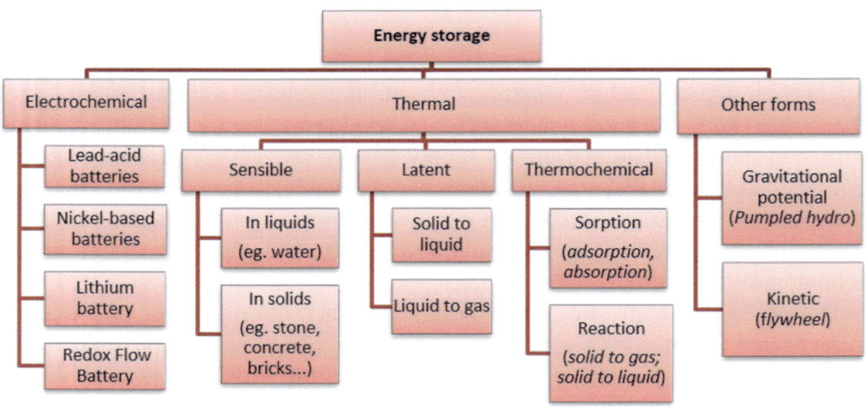

Fig. 5.1 Energy storage forms

5.2.1 Electrochemical Energy Storage

Energy stored in rechargeable electrochemical devices convert chemical potential energy into electric energy. When the terminals of the cell are loaded, between two electrodes submerged in an electrolyte within a simple cell the chemical reaction starts. Electrons are moved throughout from one electrode to the next by means of an external electric circuit or load [3]. Batteries are a beneficial electrochemical energy storage technology that stand out for their quick response times and flexible structure applied in renewable energy hybrid system [5, 6]. Batteriy life cycle, capacity, density of energy, and efficiency have all greatly increased as a result ongoing technological advancements and the prosperous creation of new battery technologies [7]. Based on the required voltage level and storage capability, a battery can have one or more cells linked either in addition, sequence, or both [3]. Three primary categories of traditional storage batteries widely utilized today are Lead–Acid Batteries, Nickel-Based Batteries, and Lithium-Based Batteries. Newer technologies such as lithium-ion and solid-state batteries are more popular due to their higher energy density, longevity and safety. Recent sodium-ion and zinc-air batteries offer cheaper and more sustainable alternatives to current battery technologies.

Lead-acid batteries: The earliest kind of rechargeable battery is lead-acid, having been developed by French physicist Gaston Plante´ in 1859. The cell power-to-weight ratio stays comparatively high because of their capacity to provide high surge currents, despite having notably low ratios of energy to weight and energy to volume. These qualities, coupled with their cost-effectiveness, make them a desirable option for use in cars, especially when it comes to the car starter motors which require supplying a substantial amount of current [4]. Lead-acid rechargeable batteries employ lead electrodes in aqueous electrolytes with sulfuric acid, making them theoretically straightforward energy storage devices. Despite their simplicity, the intricacies of the procedures for charging and discharging present challenges for performance enhancement. This technology, with a production volume of approximately 600 gigawatt-hours (GWh) and 80 billion USD revenue in 2018, holds 70% of the world market for energy storage. They are used in continuous electrical systems, the electric grid, and different motor vehicles including hybrid and lithium-ion vehicle batteries. Producers have enhanced the technology through maximizing electrode and active component performance, especially for vehicle applications, while the fundamental working principle remains unchanged. In the future, the focus is on improving battery performance by achieving quicker recharge times, extending the overall lifespan (both in terms of usage cycles and calendar life), optimizing material usage for better efficiency with active materials, and lowering the overall costs throughout the life cycle. These developments are essential for the widespread use of the grid system of storage [8].

Nickel-based batteries: which utilize nickel oxide electrodes as positive power sources, can be categorized into several types, including nickel–iron, nickel–cadmium, nickel-hydrogen, nickel-metal hydride, and nickel-zinc batteries [9].

The *nickel–iron* (Ni–Fe) *battery*, as an electrochemical source of energy that can be recharged, the battery has reached commercial maturity and is being used in Japan, Sweden, West Germany, and the USA. Inventors, Waldemar Jungner in Sweden and Thomas Alva Edison in the United States were the forerunners in creation of this kind of battery. Given the available resources and the country's climate, it turns out to be an excellent source of energy for India; however, development has not progressed to the point where commercial production is possible. Recent years have seen Karaikudi's Central Electrochemical Research Institute begin to focus intensively on the Ni–Fe system. To put it simply, the Ni–Fe battery is type of battery in which the iron electrode has been used in place of the cadmium negative [10].

The first battery utilizing a pH-neutral electrolyte was created by the Swedish scientist Waldmar Jungner in 1899, and that type of power source is called nickel–cadmium (NiCd) battery. It was a rechargeable battery with electrodes formed in a potassium hydroxide solution with nickel and cadmium. Although they are not cheap, and their cost is higher than other types of nickel-based batteries, the initial models were durable and had an energy density that was noticeably advanced than lead-acid batteries. If NiCd batteries are repeatedly emptied and refilled to the same degree of capacity, they may experience a "memory effect." The battery "remembers" when the recharging phase started in its charge cycle and experiences a sharp drop in voltage at that location during subsequent use, giving the impression that the battery has been discharged [4].

Numerous *nickel-hydrogen battery* designs exist, each with a unique benefit for certain applications. Bipolar, low-pressure metal hydride, bipolar configurations, common pressure vessel (CPV) and individual pressure vessel (IPV) are the main battery designs. Modern nickel-hydrogen batteries are mostly taking the place of nickel–cadmium batteries. Future satellites are expected to increasingly boost their power levels to more than 6 kW. Reducing the power to smaller, less expensive missions to less than 1 kW, is another trend. Therefore, lowering battery amount of volume, weight and cost is a challenge. Creating a nickel electrode that is lightweight and alternative battery designs are important. One of the existing designs, CPV nickel-hydrogen battery, is starting to show promise as an IPV design substitute. The benefits are reduced mass, volume, and manufacturing costs. A low-pressure nickel-metal-hydride battery cell has been developed, with advantage of use in aircraft [11].

In the late 1980s, Stanford R. Ovshinsky developed *the nickel-metal hydride* (NiMH) battery, a variation of the nickel–cadmium battery. In this type of battery, an alloy that absorbs hydrogen replaces the cadmium electrode used in NiCd batteries. NiMH batteries generally have a longer lifespan compared to NiCd batteries, and ongoing experimentation with new alloys continues to extend their durability. Moreover, since cadmium is toxic, NiMH batteries are more environmentally friendly [4].

In 1901, Michaelowski introduced the nickel-zinc battery through a Russian patent, and the first substantial effort to create this battery occurred in the 1930s in Ireland by Drumm. The nickel-zinc battery outperforms lead-acid and nickel–iron batteries, demonstrating specific powers exceeding 200 W kg^{-1} and specific energies ranging from 55 to 75 Wh kg^{-1}. This makes it a compelling choice for electric vehicle applications. But because of obstacles, development has advanced slowly, related to the solubility of discharge products from the zinc electrode and differences in charging efficiencies between the nickel and zinc electrodes [12].

Lithium batteries have certain advantages due to the low density of lithium and therefore provide the highest energy-to-weight ratio, and the greatest electrochemical potential. Lithium battery experiments started in 1912, and the first lithium batteries were commercialized in the 1970s. The lithium-ion power source evolved in the 1980s by a research team at Sony under the direction of American chemist John B. Goodenough, and it was a more steady and rechargeable lithium battery form. In early 90 s, the first ones went on sale [4]. One of the advantages that distinguishes this type of battery from the others are the long lifespan, high specific energy and substantial efficiency of lithium batteries. Because of these important qualities, lithium batteries are now the preferred power source for the consumer electronics industry, producing billions of units annually. Additionally, it is projected that these batteries will serve as a good candidate for electrochemical storage in renewable energy facilities and as energy sources for environmentally friendly automobiles like electric and hybrid models. Expanding the application of lithium battery technology for various uses remains challenging. There are still unresolved issues related to safety, cost, broad operating temperature ranges, and the availability of materials [13].

Redox Flow Battery: As we strive to blend sustainable energy seamlessly through the current electrical system, the purpose of energy storage becomes crucial for a cleaner, more reliable future grid. The evolving electrical grid aims to be smarter and more efficient. Redox flow battery (RFB) technology emerges as among the most Innovative fixes to address current difficulties. One kind of energy storage device that can switch between chemical and electrical energy is RFB. Usually, it consists of two redox couples that are soluble in exterior reservoirs and that are appropriate to meet certain requirements. Regenerative fuel cells, or RFBs for short, store fuel energy as dissolved redox pairs and use it to generate electricity at the electrodes. Unlike conventional batteries, which store energy in electrode materials, this is not the case. RFBs resemble solid-state batteries more due to their reversible redox reactions, even though they function somewhat similarly to conventional batteries [14].

5.2.2 Thermal Energy Storage

Thermal energy storage is a widely researched sustainable technology that addresses the challenge of balancing the fluctuating demand and supply of thermal energy from renewable sources. It stores excess energy (often in the form of heat) when electrical power is available, making it accessible for later use [15]. Thermal energy storage is a reliable way to store energy for many engineering applications including solar energy systems. The excess heat is stored over the day and used at night or during other times. Other applications of thermal energy storage are in systems for recovering waste heat and low electricity rates by accumulating the heat and releasing when there are substantial heating needs.

TES technology is always evolving and there are many commercially available systems including primarily two types: *sensible* heat storage, where energy is stored by raising the temperature of a material without changing its phase, and *latent* heat storage, which utilizes the energy absorbed or released during a material phase change, such as melting or solidification [16].

Sensible heat storage is based on high specific heat capacity medium that poses prolonged stability under thermal cycling and affordability. Sensitive heat storage includes two types, based on the storage medium used: liquid media storage (such as molten salts, oil-based fluids, water, etc.) and solid media storage (like metals, rocks, etc.) [16]. The most basic and well-established approach to storing heat is through sensible thermal energy storage (STES). Materials like water, oil, rock beds, bricks, sand, or soil fall under STES and store sensible energy by altering their temperature. In this process, when the temperature of these materials changes, there is no phase change [17]. Stored energy is determined based on the following equation:

$$Q = mC_p\Delta T$$

where, m medium mass in kg; C_p is the specific heat capacity of used medium, kJ/kg K; and ΔT is the temperature difference in K.

Latent heat storage (LHS): Latent energy storage is highly desirable because it provides a high density of energy storage. Latent heat storage is stored in Phase Change Materials (PCMs) that change their phase on constant or narrow rage temperature. For example, melting one kilogram of ice requires approximately 80 times more energy than raising the temperature of one kilogram of water by one degree Celsius. This demonstrates that a substantial amount of energy can be stored with significantly less material weight and volume when using phase change materials [16]. Commonly used materials for this purpose include water or ice, paraffin wax, and melted salt, etc. LHS can be categorized as direct or indirect based on their heat transfer methods. In an indirect system, a solid heat transfer boundary separates the heat transfer fluid, or heat transfer fluid, from the LHS material. On the other hand, in a direct system, heat transfer occurs directly without such separation [18]. Exchanged heat is determined based on the following equation:

$$\Delta Q = \Delta H = m\Delta h$$

where, m is the mass of PCM in kg; h is heat of fusion in kJ/kg.

PCMs are recognized as promising compounds for energy storage. Their ability to absorb latent heat allows them to greatly improve the energy density of buildings when compared to standard construction components [19]. Figure 5.2 shows the comparative storage capacities of different materials for sensible, latent and thermochemical heat storage highlighting the energy stored per one liter of volume.

Thermochemical energy storage relies on two main mechanisms: sorption processes and reversible chemical reactions [20]. In sorption, energy is stored through

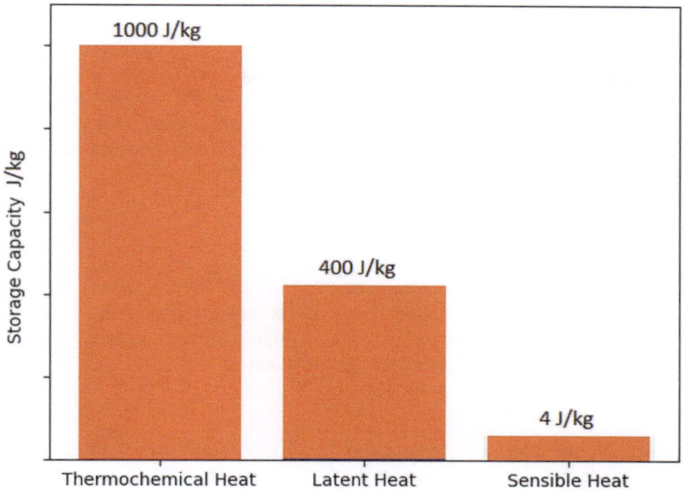

Fig. 5.2 Energy stored per 1 L of volume as thermochemical, latent and sensible

either adsorption, where molecules adhere to a surface, or absorption, where they are taken up into the bulk of a material. In reversible chemical reactions, energy is stored as the heat of reaction, which can be released when the reaction is reversed [21]. This method offers significant advantages, including high energy density and minimal thermal losses, making it well-suited for efficient, long-duration energy storage [22]. The storage density is roughly 8–10 times greater than sensitive heat storage materials and about twice as high as that of latent heat storage systems when comparing equivalent storage volumes [23]. The research in this field is in an early stage and there are no thermochemical heat storage commercial solutions for building applications [24]. Therefore, a lot of research effort must be made to improve efficiency and optimize the energy system by applying suitable algorithm [25]. Table 5.1 outlines the primary benefits and drawbacks associated with three distinct types of heat storage systems.

5.2.3 Other Forms of Energy Storage

Other forms of energy storage, such as flywheel and pumped hydro, provide efficient solutions for large-scale and long-duration energy storage. The flywheel offers high efficiency, long lifespan, and minimal environmental impact, making it suitable for applications with frequently charge–discharge cycles. Pumped hydro is a well-established technology, helps stabilize the electrical grid by storing excess energy and redistributing it during peak demand. Both systems are essential in integrating renewable energy sources and improving grid stability.

Table 5.1 Pros and cons of thermal energy storage types

Energy storage	Advantages	Disadvantages	Reference
Sensible heat	Require insulation; long lifespan; minimal environmental impact; ease of transportation due to short distances; straightforward technology; inexpensive material costs	Poor capability; significant long-term energy losses; low operating temperature	[16, 17]
Latent heat	Long-term energy losses are reasonable; there is no environmental impact; transportation is efficient due to short distances; simple technology	Poor performance; low operating temperature; limited lifespan, expensive material	[16, 18]
Thermochemical heat	Large capacity; no need for insulation; long lifespan; inexpensive material costs	Energy losses in sensible heat; high working temperature; occasionally hazardous environmental effects; poor transportation capacity due to distance; complex technology	[20, 21, 23]

Flywheel storage: When contrasted with other forms of technology, flywheel energy storage system (FESS) has a lot of benefits as a power storage option. They are famous for their long operational lifespan, megajoule (MJ) energy storage capacity, high round-trip efficiency, high power density, long life expectancy and low environmental impact. Furthermore, the total quantity of energy that can be stored in banks is infinitely large [26]. In contrast to batteries, FESS is safe in the natural world and can be used for multiple cycles of charging and discharging for a brief or long period of time. It has uses in electricity systems, supporting the incorporation of green energy sources, enhancing strength, and improving the reliability of electricity. This storage has been applied in diverse research areas, serving purposes like enhancing the reliability of electricity, supporting Uninterruptible Power Supply (UPS) systems, supporting armed forces and airplanes initiatives, preserving networks, incorporating environmentally friendly resources, boosting Flexible AC Transmission Systems (FACTS), decreasing electrical variations and supplying electricity to hybrid cars. There are two kinds of flywheel power storage: FESSs with low and high speeds. Low-speed FESS uses steel flywheels and mechanical bearings, whilst high-speed FESS uses combined disks, vacuum barriers, and magnetized bearings. It is predicted for the future that this energy storage technology will take place of batteries in numerous usage assets [27].

Pumped Hydro Energy Storage (*PHES*) has been a dependable and extensively used technique for the deep storage of electricity, since its introduction in the 1890s. Because hydro power is renewable, sustainable, and adaptable, it is used to include other intermittent sources of energy that are sustainable, such as wind and solar power, into the grid and eventually improve grid stability. Reviving obsolete small hydro power plants is becoming more popular worldwide, which emphasizes the significance of PHES. In the deregulated energy market, new strategies are necessary to optimize storage capacity and maximize PHES's profitability. Improvements in plant performance and flexibility also require advancements in turbine design. With the storage of wind and solar energy, PHES has reemerged as a technologically and financially feasible means of lowering peak load and guaranteeing power quality assurance in the early 2000s. This comeback is especially significant because large-scale electricity storage is necessary to regulate and sustain a consistent power supply in the face of unpredictable sustainable energy sources [28]. They are especially useful in stationary tasks like peak shaving and load leveling. The PHES is a reverse-engineered version of a traditional hydroelectric generation facility. A reservoir holds water, which is then released into a turbine. In a conventional hydroelectric power plant to transform the water's gravitational potential energy into electrical energy. This is where the stored water has a one-time use limit. Nonetheless, PHES offers a turbine/pumping hall and two reservoirs, one above the other [29]. Pumped hydroelectric storage stands out as the most popular technology for large-scale electricity storage. It is not only the most developed but also widely accessible commercially for bulk electricity storage needs. PHES plays a crucial role in stabilizing the electrical grid through activities like reserve generation, frequency regulation, load balancing, and peak shaving. PHES facilities fall into two main categories. The first one is pure or off-stream PHES, which gets all its energy from previously pumped water in an upper reservoir. They are also known as "closed loop" systems. The second category is pump-back, hybrid, or combined PHES, that uses energy from both naturally occurring water flowing in a stream and water that has been pumped [30].

Pumped hydro storage remains the leading technology in terms of installed capacity among energy storage power plants. This includes a large storage capacity, durability, cheap price, and quick response time. Conventional PHES outperforms Underground PHES in terms of economy and adverse environmental effects. However, because underground PHES eliminates the need for backfilling and excavation phases, it performs better in terms of social sustainability impact [31]. A system that intelligently combines two forms of storage, pumped hydro and batteries, to make it dependable, effective, and able to manage the fluctuations of renewable energy sources without consuming a lot of energy is under research [32].

5.3 Advancements in Energy Storage

It has been observed that nearly all energy storage technologies are continuously improving. These advancements have been driven by the rise of renewable energy, underscoring the growing need for efficient storage solutions, even for non-renewable green energy sources [33]. All storage technologies have certain constraints that limit their performance and improvements [34].

5.3.1 Sustainability Options

To select suitable storage options sustainability aspect must be considered [30]. Mechanical storage is the foundation of mechanical energy storage principles that include either "pressure energy" storage or kinetic and potential energy storage [35]. PHES plants historically have been serving as major as system-balancing due to their large generation capacity and cost-effectiveness. From a sustainability standpoint, PHES is notable for its long operational life, low environmental impact during operation, and ability to integrate seamlessly with renewable energy sources [36].

TES systems store excess energy, such as heat, for later use, reducing reliance on fossil fuels and minimizing energy waste. This is a key technology in supporting renewable energy integration and advancing sustainability [37]. PCMs offer high energy density and low environmental impact when using non-toxic and abundant substances. Sensible heat storage is cost-effective and utilizes widely available materials like molten salts and water, promoting low-carbon solutions for large-scale energy storage. Thermochemical energy storage, through reversible chemical reactions releases energy. The material often undergoes a complete transformation upon its discharge of chemical power. For long-term large-scale power storage, chemical energy storage is especially well-suited [38].

Fuel cells, supercapacitors, batteries, and other devices fall under electrochemical storage. These storage systems find applications in different power grid scenarios and serve as a basis for standalone power supply. In terms of sustainability, electrochemical storage supports the decarbonization of the energy sector by use of electricity from renewable sources [39].

5.3.2 Application Area

The applications of EST have a wide domain including transportation, power generation, and heating and cooling systems. Each of these mentioned requires a suitable storage solution tailored to meet performance, efficiency, and sustainability needs of particular user. Shows the main application areas of application of energy storage technologies highlighting the specific storage forms used in each sector.

Referring to Table 5.2 it is observed that EST is used in diverse applications across sectors. In transportation, chemical, hydrogen, and battery storage power electric and hybrid vehicles. Power generation relies on flywheel, pumped, magnetic, and battery storage to stabilize the grid and support renewables. For heating and cooling, thermal and thermochemical storage enhance energy efficiency in both residential and industrial settings. These varied uses underscore the critical role of EST in advancing sustainable energy systems.

Field of application is one of the criteria to determine which method of storing this energy is best for a given scenario because there are many options available and each has unique qualities. It is not possible to store energy in a way that works in

Table 5.2 Application of energy storage technologies

Area of application of EST	Storage form	Reference
Transportation	Chemical storage, hydrogen storage, battery storage	[3, 6, 7, 37]
Power generation	Flywheel storage, pumped storage, magnetic storage, hydrogen storage, battery storage	[3, 7, 6, 28–30]
Heating and cooling	Thermal storage, thermochemical storage	[15–18, 20, 21, 23]

every circumstance. Thus, they examine various energy storage methods and produce tables that highlight their advantages and disadvantages. These tables make it easier to ascertain what approach would be most beneficial in a particular circumstance. Every technology has advantages and can be applied to specific tasks (fulfilling power systems' technical requirements). However, many of these technologies are currently too costly or underdeveloped for widespread use in our current energy systems and markets, apart from Pumped Hydro Storage (PHS). It is analogous to have many different technologies in the large network of power systems, some of which are still too costly or not yet ready for widespread use [40]. The "performance index," which is the storage technique's suitability for the given application while accounting for technological features, is the basis for the comparison [41].

Energy efficiency is among the most crucial things to take into account when accurately evaluating an energy storage system, as it directly affects system performance and indirectly affects system expenses. Frequency stability, fault management, traction regenerative braking, dynamic voltage control, grid stability and a decrease in power oscillation in green energy sources are just a few of the practical applications for these kinds of energy storage technologies. A Wave Energy Converter generation scenario has been selected as the study case for the use of compensating oscillations of electricity in the production of renewable energy [41, 42].

Investment cost is an important decision factor as well, e.g. Pumped Heat Energy Storage is an emerging storage technology that stores energy in the form of heat. It works by using electricity to drive a heat pump, which moves heat from one location to another. The stored energy is in the form of thermal energy, and it can later be converted back into electricity using a heat engine or turbine when needed. Pumped Heat Energy Storage stores heat at various temperatures by using two energy storage tanks, compressed and expanded gas. The price of storing electrical power with this type of technology is comparable to calculating the cost of energy conservation. Depending on certain assumptions, the cost for a test plant ranged from 2.7 to 5.0 Euro cents per kilowatt-hour. This Pumped Heat Energy Storage was just as economical when compared to other techniques such as Compressed Air Storage. It might even be competitive with pumped hydro storage, which also offers the advantage of more placement flexibility. However, just like any other tech item, the price is based on how frequently it is used. Even though pumped heat energy storage was designed for short-term use, it looks like a promotable option for long-term storage. Nevertheless, additional empirical data and a thorough cost analysis are necessary to confirm its cost-effectiveness [43].

Volume or mass density play an important role in energy storage technologies. Flexible power sources are becoming more and more necessary as wearable electronics continue to grow quickly. Flexible rechargeable batteries possess an advanced level of flexibility, a stable output, and a density of energy that makes them essential for energy storage in flexible systems. Nevertheless, extreme flexibility frequently comes with a price of energy density. Considering the advancements in flexible power storing devices, a thorough evaluation of their energy density is imperative [44]. When deciding on an energy storage method, there are two key elements to consider:

- *Mass density* is crucial because the cost depend so material weight, especially for permanent storage. To reduce expenses, it is preferable to used material that stores a reasonable amount of energy, rather than the maximum [41].
- *Volume* matters when storage must fit in a limited or expensive space, or when size affects systems like cooling. Larger volumes need more materials and increase construction costs [41].

Therefore, choosing the best storage option includes opting between higher energy density in a smaller space or accepting a larger volume to keep overall costs lower [41].

5.4 Implementation Challenges and Opportunities

Renewables like wind and solar are displacing traditional energy generation to distributed generation. Energy storage as one of major components of sustainable energy systems faces several challenges including technical, economic, and regulatory. At the same time these challenges are opportunities for the development of more efficient and cost-effective solutions and opportunities for economic growth.

5.4.1 Challenges

Energy storage systems are enabler technology for demand-side management and distributed generation. Energy storage installation near users is required for boosting local renewable energy generation utilization. Energy storage can contribute to decarbonizing transportation, heating, and electricity sectors, lower greenhouse gas emissions [47]. These systems open opportunities for efficient energy demand management and help lower carbon emissions in the transportation, heating, and electricity industries [45, 46]. Innovative technologies and alternative energy sources are driving a change in this field. Energy storage can decarbonize transportation, heating, and electricity sectors, lower greenhouse gas emissions and advancing sustainability [47].

The integration of heat pumps and thermal energy storage systems presents challenges in incorporating substantial amounts of electricity into the grid. Effective operation of these systems, particularly during periods of high electricity demand, is contingent upon the coordinated use of thermal storage and the flexible operation of heat pumps. Although hot water storage and heat pumps help balance renewable energy variations and manage electricity demand, their efficiency may be impacted by the price of electricity [48]. Thermal management can be improved even in the most basic buildings by utilizing materials such as *phase change materials*. Nevertheless, the advantages of these systems might be undermined by the conventional half-hourly electricity pricing mechanism, especially when starting from room temperature. To enable demand-side management and balance the variability of renewable sources generation, this suggests that a new energy paradigm that combines these technologies with communication and control strategies is required. A lot of research has been done on this topic related to challenges of power storage technologies [49].

Technology challenges for energy storage must advance in terms of capacity, durability, affordability, and security. To enable its practical deployment, research efforts are focused on simulating and optimizing energy storage across a range of applications. Research, development, and practical utilization of energy storage tanks would become supported by the establishment of comprehensive standards and increased transparency. Renewable energy storage has become more competitive due to the significant decrease in costs for energy storage systems. To put it simply, this storage has an important function in both integrating sustainable energy sources and overcoming obstacles in the energy sector [49].

Economic challenges include the development of energy storage solutions that will require involvement of a variety of stakeholders including energy users, electrical enterprises, researchers, and social organizations. This will guarantee the incorporation of sustainable development models and assist in the establishment of renewable energy sources [49].

Microgrid applications are good examples related to economic challenges. Energy storage system (ESS) quality management system (QMS) implementation in microgrid (MG) applications can streamline operations, boost overall efficiency, and cut expenses. A QMS facilitates risk management, expedites problem solving, lowers operational errors, and enhances workflow. A QMS contributes to the consistent and dependable performance of ESS in MG applications by guaranteeing that quality standards are met at every stage of the execution. Performance gains, increased client satisfaction, and a more economical energy management system are the results of this [50]. Table 5.3 outlines the key challenges and opportunities associated with energy storage implementation, categorized across technical, economic, regulatory, and social dimensions.

Cost and technical challenges associated with energy storage are defined by the expense of storage energy capacity. The technical feasibility obstacles are assessed through the minimum power-to-energy ratio of storage, in relation to its energy capacity. The ability of storage to compete with reliable low carbon generating sources significantly relies on the costs of storage energy capacity and its discharge

Table 5.3 Categorized challenges and opportunities in energy storage implementation

Category	Challenges	Opportunities	References
Technical and economic	• High capital costs and uncertain ROI • Limited durability, safety, and performance of some storage technologies	• Declining lithium-ion battery costs • Demand-side anagement and local renewable energy optimization	[51, 52]
Grid and technology	• Complex technology selection • Integration with heat pumps and demand variability • Grid instability during peak demand periods	• Tailored tech: pumped hydro, compressed air, batteries • Enhances grid stability and resilience	[53–55]
Policy and regulation	• Regulatory uncertainty • Lack of standardized frameworks • Legal/institutional barriers	• Supportive policies in US, UK, China • Subsidies and green mandates	[53, 55]
Social and research	• Public opposition to infrastructure • Low awareness of ESS benefits • Fragmented research landscape	• Job creation and sustainability alignment • Innovation in modeling, control, and materials	[52, 56, 57]

efficiency. Given the substantial renewable energy generation capacities and ambitious electrification targets in the ZEB 2050 scenario, new strategies are being explored to implement the most economically advantageous storage solution [72]. Research indicates that affordable energy capacity combined with extended storage technologies plays a crucial role in system integration. Over time, the investment costs for electricity storage systems are expected to decrease significantly, particularly during the 2030s. The average annual prices for batteries used in stationary applications are projected to decline notably in the medium term. As battery prices become more competitive for grid storage, this is likely to encourage an increase in electricity storage system installations in the coming years. The optimal level of electricity storage capacity is anticipated to reach a substantial figure in the 2040s within the baseline ZEB 2050 scenario that includes affordability targets. This level corresponds to a significant proportion of the total installed wind energy capacity. Financial support for ZEB 2050 to implement electricity storage solutions from a building standpoint is essential during every subsidy period [72].

5.4.2 *Opportunities*

The advancement of technologies for energy storage creates new job opportunities and calls for significant engineering investments. It can be difficult to choose the best technology for every application with so many options available. But as costs come down and technology advances, energy storage is growing, making it more sustainable and efficient [58]. Some nations, like USA, Japan, China, United Kingdom etc. have suggested implementing these technologies policies to be able to facilitate integration of renewable energy sources and maintain grid stability. These regulations concentrate on batteries, which are becoming more and more flexible, economical, and efficient. The goals of ESS policies are to speed up the switch to clean energy, balance supply and demand, and enhance local consumption of renewable energy. They are mostly found in developed economies with a strong energy sector [59]. As an example of energy storage opportunities can be related to mine shaft. The hoisting machinery, currently available, is use only 40–50% of the time. By boosting machine uptime, energy storage in mine shafts can raise mine profitability. But to do this, organizational and legal regulations must be changed. With these adjustments, mines would be able to utilize energy storage, which would boost both productivity and effectiveness [60].

Regulators and policy makers around the world face disruptive implications from energy storage technologies, whether they are deployed at large-scale installations or dispersed microscale installations. To meet decarbonization targets and commitments under the Paris Agreement while preserving the security and stability of the energy supply, large-scale battery storage can effectively increase the variable renewable sources involvement like solar and wind. Nevertheless, there are several obstacles to overcome when incorporating massive amounts of energy storage from batteries into the electrical network, including social, technological, policy, and scientific ones. The goal of ongoing research in large-scale battery technology is to improve efficiency, reliability, and the shift to renewable energy generation and storage technologies that are more environmentally friendly and globally efficient [61].

5.5 Trends and Research Directions in Energy Storage Systems

In recent years, the field of energy storage technologies have experienced significant growth, driven by the increasing demand for efficient and sustainable energy solutions. This surge in interest is reflected in the growing number of patents filed over the past two decades, which highlights the ongoing innovations and advancements in the sector. Figure 5.3 provides a visual representation of the number of patents in energy storage technologies, offering insights into key trends and emerging research directions in this rapidly evolving field.

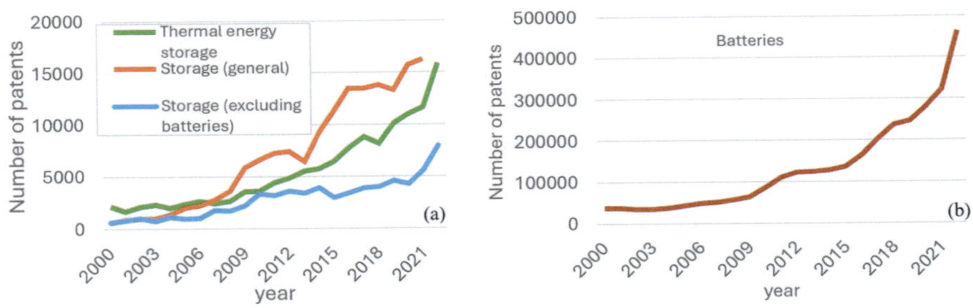

Fig. 5.3 Global annual patent applications in energy storage technologies: **a** thermal storage; **b** batteries (The author: based on data from [62])

Costs, energy density, specific power, durability, recyclability, energy efficiency, and other aspects are a set of evaluating criteria how well various storage systems perform [41]. Every energy storage technology has certain special qualities that make it appropriate for a certain type of energy storage application. These distinctive qualities aid in determining which energy storage technology is most appropriate to use in each situation [63].

Therefore, implementing energy storage systems in buildings can contribute to energy efficiency, reduce carbon emissions, and provide opportunities for sustainable energy consumption. Consideration of factors such as system design, climatic conditions, and the specific type of ESS used is important in maximizing the benefits of these systems. It is crucial to consider the hybrid energy storing systems (HESS) integrated with energy systems that are sustainable. Power system operators, researchers, developers, and manufacturers can in collaboration create HESSs that are efficient, affordable, and contribute to green energy solutions [64].

Significant advances are currently being made in battery technology. Scientists are investigating solid-state and sodium-ion battery technologies, creating new materials, and refining lithium-ion batteries. To be able to fulfill the growing need for power storage across a variety of sectors, these advancements seek to improve energy storage, resilience, and safety. While some batteries are still in the experimental stage, others are commercially available [65]. The lithium-ion rechargeable batteries, currently used in powered automobiles (EV), have power storage capability and safety that are limited. The goal for EV manufacturers is to lower the cost of their vehicles without sacrificing functionality or range. To overcome these obstacles and advance battery technology, research is still being done. As more people switch to powered vehicles due to environmental concerns, efforts are being made to increase battery sustainability [66]. Given its lengthy development cycles and high cost, it is unlikely that alternative technologies will displace lithium-ion battery technology within the next ten years. Proper pack design is essential for safety and longevity, and different battery technologies call for different management systems [67]. Major challenges and research direction on Energy Storage System are summarized in Table 5.4.

Table 5.4 Challenges and associated research directions for energy storage systems

Challenges of EST	Future directions	Reference
Advanced battery technologies	Solid-state, sodium-ion, lithium-ion, resilience, safety, sustainability, development cycles, environmental concerns, electric vehicles (EVs), cost reduction, functionality, experimental stage, management systems	[65–68]
Energy storage for renewables integration	Grid integration, renewable energy, battery storage, smart grids, energy management, storage technologies, peak shaving, dispatchability, hybrid systems, microgrids Storage capacity	[69–71]
Next-generation thermal storage	High-temperature storage, phase change materials, molten salt, energy efficiency, thermal energy, thermal insulation, sensible heat, high-capacity storage, latent heat, renewable energy integration, thermochemical storage	[72, 73]
Combining energy storage technologies	Combined storage systems, integrated energy storage, hybridization, dual-storage systems, enhanced efficiency, increased reliability, improved flexibility, scalability	[74, 75]

With the development of advanced capacitors for improved power quality, flywheel energy supply technologies, and superconducting power storage systems, battery technology has improved to be used in electric vehicles and utility applications. Performance advantages can be obtained in utility applications by integrating energy storage into electronic compensators. The growing family of custom power, power quality, and flexible ac transmission systems (FACTS) is well-suited to energy storage technology. The stability, transmission capacity, and quality of energy of electric systems can be improved by integrating energy storage with power electronics converters [68].

To create a dependable and sustainable power grid, researchers are tackling problems such as extending the life of batteries, guaranteeing energy efficiency, sizing power system of storage appropriately, and evaluating whether the cost of these systems justifies their purchase. ESS integration with the electrical grid can be difficult. Researchers recommend overcoming the following obstacles to ensure a successful ESS integration: battery ESS degradation, energy efficiency, ESS sizing and allocation, and ESS financial viability [69].

Comparatively small storage capacities can lead to notable reductions in power imbalance, The power imbalance brought on by renewable energy unpredictability can be lowered using energy storage. The power imbalance problem is formulated as a stochastic control problem using an analytical model that considers various timescales of grid operation. It demonstrates how the average power imbalance can be reduced by an avaricious policy [70]. The difficulties in assessing energy storage's place in the changing energy landscape is important. It highlights how important it is to develop new models that can shed light on the technical, financial, and legal ramifications of storage integration. It also recognizes the obstacles to the widespread use of energy storage, including issues with cycle efficiency, cost, and technological maturity. Argues that demand response and energy curtailment are two potential pairings of resources to consider in addition to energy storage as a comprehensive answer for the problems connected with integration of renewable energy. In general, the significance of tackling technical issues as well as legal and market obstacles when integrating energy storage [71].

Based on the literature a next generation thermal energy storage is discussed in a way to store heat at very high temperatures with an inexpensive, stable material. The expenses associated with storing and utilizing this heat are expected be reduced [72]. Since 2010, considerable progress has occurred in the investigation and creation of next generation Concentrated Solar Power (CSP). This advancement centered on the exploration of innovative heat energy storage aiming at higher temperatures and more efficient power cycles. The four primary concepts investigated included solid particles, specific molten salts (especially chlorides), gases utilizing indirect thermal storage, and liquid metals with diverse TES options. While these technologies offer advantages for CSP plants, there are challenges that must be addressed, such as tackling corrosion issues and optimizing system efficiency [73].

According to research done by Power Electronics Society in an Institution of Electrical Engineers there are three dependable energy storage systems that functioned creatively, combined in a novel way by them. It is anticipated that in the near future, this novel approach will lead to technically and economically feasible battery substitutes. The end products might provide better performance and efficiency while addressing the drawbacks of conventional battery technologies. This innovation represents a major advancement in energy storage and has potential uses in several fields. Active Power's TEX technology distinguishes itself from other energy storage technologies by inventively fusing three established technologies. By overcoming these obstacles, the integration promises advantages over conventional battery technologies, including increased capacity, efficiency, cycle life, and possibly lower costs. Utilizing the advantages of several different technologies, TEX offers an adaptable and durable solution for energy storage applications [74]. Through the hybrid approach, or HESS, critical metrics like cycle life, efficiency, and energy storage capacity can be improved by combining various components with different properties. The hybrid strategy possesses the power to transform storing energy and advance cause of more ecologically harmless energy sources in the future [75].

5.6 Concluding Remarks

Energy storage technologies play a crucial role in the transition to sustainable and efficient energy production, especially in the context of ZEB. As ZEBs aim to produce as much energy as they consume, among the others they rely on energy storage solutions to balance the supply and demand. Despite challenges such as high capital costs, regulatory uncertainties, and integration complexities, advancements in energy storage solutions are still significant. The increasing focus on sustainability, coupled with innovations in materials and hybrid systems, is opening up new opportunities for more flexible and cost-effective energy storage solutions. Global efforts to address these challenges through research, technology development, and regulatory support are essential for overcoming barriers and ensuring that ESS can contribute to decarbonizing energy systems.

References

1. Rahman MM, Oni AO, Gemechu E, Kumar A (2020) Assessment of energy storage technologies: a review. https://doi.org/10.1016/j.enconman.2020.113295
2. Chen H, Cong TN, Yang W, Tan C, Li Y, Ding Y (2009) Progress in electrical energy storage system: a critical review. Science Press. https://doi.org/10.1016/j.pnsc.2008.07.014
3. Hadjipaschalis I, Poullikkas A, Efthimiou V (2009) Overview of current and future energy storage technologies for electric power applications. https://doi.org/10.1016/j.rser.2008.09.028
4. Wagner L (2013) Overview of energy storage technologies. In: Future energy: improved, sustainable and clean options for our planet, pp 613–631. https://doi.org/10.1016/B978-0-08-099424-6.00027-2
5. Abed FT, Abed NK, ALRubeei IRN, A lrikabi (204) Using genetic algorithm for optimal sizing of stand-alone hybrid energy system. Herit Sustain Dev 6(1):287–300. https://doi.org/10.37868/HSD.V6I1.501
6. Bradbury, K "Energy Storage Technology Review," 2010 [Online]. Available: https://www.researchgate.net/publication/265064873
7. Zhang C, Wei YL, Cao, Lin MC (2018) Energy storage system: current studies on batteries and power condition system. https://doi.org/10.1016/j.rser.2017.10.030
8. Lopes PP, Stamenkovic VR (2020) Past, present, and future of lead–acid batteries. American Association for the Advancement of Science. https://doi.org/10.1126/science.abd3352
9. Shukla AK, Venugopalan S, Hariprakash B Nickel-based rechargeable batteries
10. Chakkaravarthy C, Periasamy P, Jegannathan S, Vasu KI (1991) The nickel/Iron battery
11. Smithrick JJ, O'Donnell PM (1996) Nickel-hydrogen batteries—an overview. J Propuls Power 12(5):873–878. https://doi.org/10.2514/3.24116
12. Mcbreen J (1994) Nickel/zinc batteries
13. Scrosati B, Garche J (2010) Lithium batteries: status, prospects and future. https://doi.org/10.1016/j.jpowsour.2009.11.048
14. Wang W, Luo Q, Li B, Wei X, Li L, Yang Z (2013) Recent progress in redox flow battery research and development. Adv Funct Mater 23(8):970–986. https://doi.org/10.1002/adfm.201200694
15. Lefebvre D, Tezel FH (2017) A review of energy storage technologies with a focus on adsorption thermal energy storage processes for heating applications. https://doi.org/10.1016/j.rser.2016.08.019.
16. Hasnain SM Review on sustainable thermal energy storage technologies, part I: heat storage materials and techniques
17. Koçak B, Fernandez AI, Paksoy H (2020) Review on sensible thermal energy storage for industrial solar applications and sustainability aspects. https://doi.org/10.1016/j.solener.2020.08.081
18. Jouhara H, Żabnieńska-Góra A, Khordehgah N, Ahmad D, Lipinski T (2020) Latent thermal energy storage technologies and applications: a review. Int J Thermofluids 5–6. https://doi.org/10.1016/j.ijft.2020.100039
19. Sarbu I, Sebarchievici C (2018) A comprehensive review of thermal energy storage. MDPI. https://doi.org/10.3390/su10010191

20. Carrillo AJ, González-Aguilar J, Romero M, Coronado JM (2019) Solar energy on demand: a review on high temperature thermochemical heat storage systems and materials. Chem Rev 119(7):4777–4816. https://doi.org/10.1021/acs.chemrev.8b00315
21. Pardo P, Deydier A, Anxionnaz-Minvielle Z, Rougé S, Cabassud M, Cognet P (2014) A review on high temperature thermochemical heat energy storage. https://doi.org/10.1016/j.rser.2013.12.014
22. Duraković B (2020) Introduction. In: Green energy and technology, pp 1–15. https://doi.org/10.1007/978-3-030-38335-0_1
23. Aydin D, Casey SP, Riffat S (2015) The latest advancements on thermochemical heat storage systems. https://doi.org/10.1016/j.rser.2014.08.054
24. Tatsidjodoung P, Le Pierrès N, Luo L (2013) A review of potential materials for thermal energy storage in building applications Author links open overlay panel. Renew Sustain Energy Rev 18: 327–349. https://doi.org/10.1016/j.rser.2012.10.025
25. Siasiev A, Dudnikov S, Nikolaienko D, Hubal H, Bondarchuk O (2025) Evaluation of the effectiveness of decision-making algorithms in the management of renewable energy sources: a case study of Ukraine. Sustain Eng Innov 7(1):87–96. https://doi.org/10.37868/SEI.V7I1.ID415
26. Amiryar ME, Pullen KR (2017) A review of flywheel energy storage system technologies and their applications. Balk Soc Geometers. https://doi.org/10.3390/app7030286
27. Arani AAK, Karami H, Gharehpetian GB, Hejazi MSA (2017) Review of flywheel energy storage systems structures and applications in power systems and microgrids. https://doi.org/10.1016/j.rser.2016.11.166
28. Rehman S, Al-Hadhrami LM, Alam MM (2015) Pumped hydro energy storage system: a technological review. https://doi.org/10.1016/j.rser.2014.12.040
29. Hossain E, Faruque HMR, Sunny MSH, Mohammad N, Nawar N (2020) A comprehensive review on energy storage systems: types, comparison, current scenario, applications, barriers, and potential solutions, policies, and future prospects. MDPI AG. https://doi.org/10.3390/en13143651
30. Yang CJ (2016) Pumped hydroelectric storage. Storing Energy 25–38, With special reference to renewable energy sources. https://doi.org/10.1016/B978-0-12-803440-8.00002-6
31. Guo Z, Ge S, Yao X, Li H, Li X (2020) Life cycle sustainability assessment of pumped hydro energy storage. Int J Energy Res 44(1):192–204. https://doi.org/10.1002/er.4890
32. Javed MS, Zhong D, Ma T, Song A, Ahmed S (2020) Hybrid pumped hydro and battery storage for renewable energy based power supply system. Appl Energy 257. https://doi.org/10.1016/j.apenergy.2019.114026
33. Sabihuddin S, Kiprakis AE, Mueller M (2015) A numerical and graphical review of energy storage technologies. MDPI AG. https://doi.org/10.3390/en8010172
34. Baker J (2008) New technology and possible advances in energy storage. Energy Policy 36(12):4368–4373. https://doi.org/10.1016/j.enpol.2008.09.040
35. Sterner M, Stadler I Handbook of energy storage
36. Castagneto Gissey G, Dodds PE, Radcliffe J (2018) Market and regulatory barriers to electrical energy storage innovation. https://doi.org/10.1016/j.rser.2017.09.079
37. ALRubeei IRN, Idi SN, Hussein Alsammak IL, AlRikabi HT, Mutar HA, Alaidi AHM (2025) Using artificial intelligence for enhancement of solar cell efficiency in the of Iraq. Herit Sustain Dev 7(1):197–210. https://doi.org/10.37868/HSD.V7I1.1067
38. Guney MS, Tepe Y (2017) Classification and assessment of energy storage systems. https://doi.org/10.1016/j.rser.2016.11.102
39. Fernão Pires V, Romero-Cadaval E, Vinnikov D, Roasto I, Martins JF (2014) Power converter interfaces for electrochemical energy storage systems—a review. Energy Convers Manag 86:453–475. https://doi.org/10.1016/j.enconman.2014.05.003
40. Luo X, Wang J, Dooner M, Clarke J (2015) Overview of current development in electrical energy storage technologies and the application potential in power system operation. Appl Energy 137:511–536. https://doi.org/10.1016/j.apenergy.2014.09.081

41. Ibrahim H, Ilinca A, Perron J (2008) Energy storage systems—characteristics and comparisons. https://doi.org/10.1016/j.rser.2007.01.023
42. Torres J, Moreno-Torres P, Navarro G, Blanco M, Lafoz M (2018) Fast energy storage systems comparison in terms of energy efficiency for a specific application. IEEE Access 40656–40672. Institute of Electrical and Electronics Engineers Inc. https://doi.org/10.1109/ACCESS.2018.2854915
43. Smallbone A, Jülch V, Wardle R, Roskilly AP (2017) Levelised cost of storage for pumped heat energy storage in comparison with other energy storage technologies. Energy Convers Manag 152:221–228. https://doi.org/10.1016/j.enconman.2017.09.047
44. Wang D et al (2020) Energy density issues of flexible energy storage devices. Energy Storage Mater 28:264–292. https://doi.org/10.1016/j.ensm.2020.03.006
45. Parra D et al (2017) An interdisciplinary review of energy storage for communities: challenges and perspectives. https://doi.org/10.1016/j.rser.2017.05.003
46. Jain R et al (2023) Optimization of energy consumption in smart homes using firefly algorithm and deep neural networks. Sustain Eng Innov 5(2):161–176. https://doi.org/10.37868/SEI.V5I2.ID210
47. AL Shaqsi AZ, Sopian K, Al-Hinai A (2020) Review of energy storage services, applications, limitations, and benefits. https://doi.org/10.1016/j.egyr.2020.07.028
48. Hewitt NJ (2012) Heat pumps and energy storage—the challenges of implementation. Appl Energy 89(1):37–44. https://doi.org/10.1016/j.apenergy.2010.12.028
49. Yao L, Yang B, Cui H, Zhuang J, Ye J, Xue J (2016) Challenges and progresses of energy storage technology and its application in power systems. J Mod Power Syst Clean Energy 4(4):519–528. https://doi.org/10.1007/s40565-016-0248-x
50. Faisal M, Hannan MA, Ker PJ, Hussain A, Bin Mansor M, Blaabjerg F (2018) Review of energy storage system technologies in microgrid applications: issues and challenges. Institute of Electrical and Electronics Engineers Inc. https://doi.org/10.1109/ACCESS.2018.2841407
51. Mongird K et al (2020) An evaluation of energy storage cost and performance characteristics. Energies 2020 13(13):3307. https://doi.org/10.3390/EN13133307
52. Opazo MA (2023) Climate crisis: energy storage challenges in the transition to renewable energies. RSC Sustainability 1(7):1602–1603. https://doi.org/10.1039/D3SU90038J
53. Yao L, Yang B, Cui H, Zhuang J, Ye J, Xue J (2016) Challenges and progresses of energy storage technology and its application in power systems. J Mod Power Syst Clean Energy 4(4):519–528. https://doi.org/10.1007/S40565-016-0248-X/FIGURES/4
54. Rana MM et al (2023) Applications of energy storage systems in power grids with and without renewable energy integration—a comprehensive review. J Energy Storage 68:107811. https://doi.org/10.1016/J.EST.2023.107811
55. Ahmed MMS et al (2024) Prospects and challenges of energy storage materials: a comprehensive review. Chem Eng J Adv 20:100657. https://doi.org/10.1016/J.CEJA.2024.100657
56. Adeyinka AM, Esan OC, Ijaola AO, Farayibi PK (2024) Advancements in hybrid energy storage systems for enhancing renewable energy-to-grid integration. Sustain Energy Res 2024 11(1):1–23. https://doi.org/10.1186/S40807-024-00120-4
57. Dolge K, Toma AS, Grāvelsiņš A, Blumberga D (2023) Realizing renewable energy storage potential in municipalities: identifying the factors that matter. Environ Clim Technol 27(1):271–288. https://doi.org/10.2478/RTUECT-2023-0021
58. Ozdemir E, Ozdemir S, Erhan K, Aktas A 3 rd International smart grid workshop and certificate program (ISGWCP) Istanbul, Turkey
59. Sani SB et al (2020) Energy storage system policies: way forward and opportunities for emerging economies. https://doi.org/10.1016/j.est.2020.101902
60. Siostrzonek T (2023) The mine shaft energy storage system—implementation threats and opportunities. Energies (Basel) 16(15). https://doi.org/10.3390/en16155615
61. Faunce TA, Prest J, Su D, Hearne SJ, Iacopi F (2018) On-grid batteries for large-scale energy storage: challenges and opportunities for policy and technology. https://doi.org/10.1557/mre.2018.11.

62. Annual patents filed for energy storage technologies, World. Accessed 04 Apr 2024 [Online]. Available: https://ourworldindata.org/grapher/patents-energy-storage?showSelectionOnlyInTable=1
63. Aneke M, Wang M (2016) Energy storage technologies and real life applications—a state of the art review. https://doi.org/10.1016/j.apenergy.2016.06.097
64. Atawi IE, Al-Shetwi AQ, Magableh AM, Albalawi OH (2023) Recent advances in hybrid energy storage system integrated renewable power generation: configuration, control, applications, and future directions. MDPI. https://doi.org/10.3390/batteries9010029
65. Divya KC, Østergaard J (2009) Battery energy storage technology for power systems—an overview. https://doi.org/10.1016/j.epsr.2008.09.017
66. Loganathan MK, Ming Tan C, Mishra B, Msagati TAM, Snyman LW (2019) Review and selection of advanced battery technologies for post 2020 era electric vehicles. In: 2019 IEEE Transportation electrification conference, ITEC-India 2019, Institute of Electrical and Electronics Engineers Inc. https://doi.org/10.1109/ITEC-India48457.2019.ITECIndia2019-254
67. Zeng X et al (2019) Commercialization of lithium battery technologies for electric vehicles. https://doi.org/10.1002/aenm.201900161
68. Ribeiro PF, Johnson BK, Crow ML, Arsoy A, Liu Y (2001) Energy storage systems for advanced power applications
69. Tan KM, Babu TS, Ramachandaramurthy VK, Kasinathan P, Solanki SG, Raveendran SK (2021) Empowering smart grid: a comprehensive review of energy storage technology and application with renewable energy integration. https://doi.org/10.1016/j.est.2021.102591
70. Su HI, El Gamal A (2013) Modeling and analysis of the role of energy storage for renewable integration: power balancing. IEEE Trans Power Syst 28(4):4109–4117. https://doi.org/10.1109/TPWRS.2013.2266667
71. Castillo A, Gayme DF (2014) Grid-scale energy storage applications in renewable energy integration: a survey. https://doi.org/10.1016/j.enconman.2014.07.063
72. Ho C et al (2014) Technology advancements for next generation falling particle receivers. Energy Procedia 398–407. https://doi.org/10.1016/j.egypro.2014.03.043
73. Ding W, Bauer T (2021) Progress in research and development of molten chloride salt technology for next generation concentrated solar power plants. https://doi.org/10.1016/j.eng.2020.06.027
74. Institution of Electrical Engineers, IEEE Power Electronics Society, INTELEC 2004 (2004) The 26th annual international telecommunications energy conference. In: Proceedings : exploring power ; the foundation for communications. IEEE, Chicago, Hyatt Regency, Chicago, Illinois, USA
75. Bocklisch T (2016) Hybrid energy storage approach for renewable energy applications. J Energy Storage 8:311–319. https://doi.org/10.1016/j.est.2016.01.004

Chapter 6
Current and Emerging Materials for Zero Energy Buildings

Abstract Construction materials have been gaining attention and popularity in recent years due to the global warming and climate change concerns. This chapter aims to analyze and review current materials being used in construction, their advantages, and disadvantages, as well as promising material that may offer viable alternatives to our conventional materials in the future. Recent recycling technologies and advancements in material science enabling a new opportunity to potentially develop innovative and eco-friendly construction materials that maintain performance and reduce environmental impact.

Keywords Sustainable construction materials · Green concrete · SBM · Biochar · Carbon nanotubes

Abbreviations

CCM	Complementary cementing materials
CCS	Carbon capture and storage
CNT	Carbon nanotubes
SMB	Stabilized mud blocks
PCMs	Phase Change Materials
ZCCM	Zero Carbon Construction Materials
ZEBs	Zero Energy Buildings

6.1 Introduction

Rapid economic growth has resulted in alarming levels of pollution and deteriorate the issue of global climate change. Notably, global CO_2 emissions have increased by more than 60% since 1990, and in 2022 reached an enormous 37.5 billion metric tons ($GtCO_2$) from fossil fuels [1]. On the other side, the looming depletion of

B. Duraković, *Zero Energy Buildings*, Green Energy and Technology,
https://doi.org/10.1007/978-3-032-16027-0_6

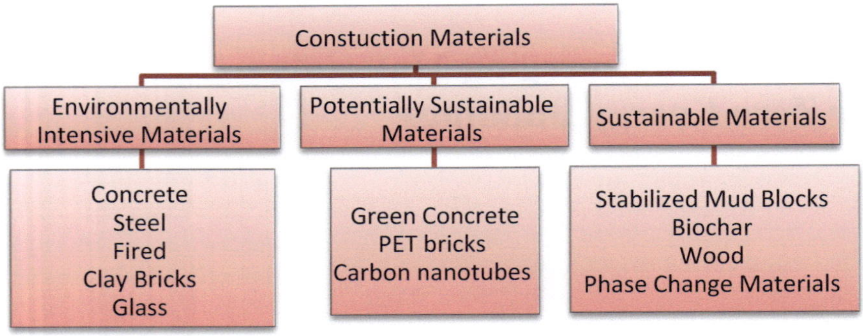

Fig. 6.1 Construction materials

fossil fuel reserves presents an imminent threat. At current rate of consumption, the depletion of fossil fuels is rapidly growing, and it is estimated that exhaustion of oil reserves by 2052, gas by 2060, and coal by 2090 [2], indicating an upcoming energy crisis. This situation along with concurrent global economic challenges and the increased depletion of non-renewable energy sources, caused increased awareness of energy conservation [3]. Over the past four decades, energy consumption has swiftly increased by an annual average of 1.8%, with trend that indicated continuation over 58% of all energy consumption by 2050 [4]. Notably, a significant portion of this energy demand comes from buildings and the construction sector, reaching its share between 32 to 40% of global energy consumption [5, 6].

Due to the facts discussed above, efforts to mitigate building energy demand have directed research into the development and utilization of construction materials driving sustainability. This chapter aims to explore these materials including their strengths and weaknesses, comparing them against current building materials, exploring integration challenges and future trends. Figure 6.1 shows some of the construction materials and their impact on the environment.

With the aim to show construction material sustainability potential, Fig. 6.1 is organized into three main categories sorting sustainable and potentially sustainable materials from environmentally intensive materials. Sustainable materials such as "zero carbon material" refers to any material, structure or activity which releases a total of net-zero carbon emissions yearly. For example, a building which runs entirely on solar power without the use of any fossil fuels would be considered as a zero-carbon or carbon neutral structure [7]. One of the best ways to achieve a zero-carbon state is to use sustainable building materials. The most sustainable building materials are those that are naturally occurring such as soil, stones, timber, and biomass. However, these materials have limitations, especially in terms of strength and durability. Processing and transportation require energy expenditure, which results in carbon emissions. To reduce carbon emissions, it will become necessary to devise technologies that produce building materials and products with the least amount of energy expenditure [8, 9]. A review of some building materials and techniques is provided in this chapter [10].

Therefore, chapter examines some selected construction materials that have been used for Zero Energy Buildings (ZEBs) by categorizing them based on sustainability: *less sustainable/unsustainable* (e.g., concrete, steel, fired brick etc.), *sustainable* (e.g., wood, stabilized mud blocks, biochar), and *potentially sustainable* materials that improve with proper use and production methods (e.g., green concrete, etc.).

6.2 Environmentally Intensive Materials

These materials have a high environmental impact due to their extraction, production, or transportation processes. They consume large amounts of energy and produce a large carbon footprint, which may require toxic chemicals for processing. As ZEBs aim to reduce environmental impact, it is important to limit the use of these materials. Table 6.1 presents an overiew of existing mainstream construction materials.

To understand the importance of construction materials driving sustainability, the current state of construction materials and how they impact our environment must be understood. There is an extensive variety of materials used in the construction sector. Different materials are employed in construction, depending on each material's structural capabilities. Local and national regulations control the building materials used in construction in various areas. Choosing the correct material for a project means enhancing and lengthening a project's lifetime. Currently mostly used construction materials are carbon-intensive material due to its production process. A summary of current materials is available at the end of this section in Table 6.1.

6.2.1 Concrete

Concrete is an artificial, stone-like composite material created by combining aggregate (such as sand, gravel, stone, or brick chips), water, admixtures, and other materials in specific, predetermined amounts with a binding material (such as cement or lime). The ratios used in the mixing determine the strength and quality [14, 15]. Cement is a carbon-intensive material due to its production process. While there are efforts to make cement more sustainable (e.g., using alternative binders), traditional cement is not considered sustainable. There are ongoing innovations to reduce the carbon footprint, but it remains a significant challenge to make traditional concrete truly sustainable. Concrete is strong, simple to make, and adaptable to a variety of sizes and forms. In addition to that, it is affordable, readily combined, and sensible. It allows for dependable, high quality fast-track construction. Concrete constructions should usually be strong enough to withstand natural disaster such as: typhoons, hurricanes, tornadoes, and earthquakes. Concrete structures have survived both wars and natural disasters, with some outlasting the very civilizations that created them

Table 6.1 Material performance and sustainability overview for common building materials

Material	Production	Cost	Functionality	Sustainability	Implementation	Ref
Concrete	Easy and economical, 30 billion tonnes produced annually	Very economical and cheap to produce	Impressive strength, can withstand natural and man-made disasters	Colossal carbon footprint, accounts for 8% of global CO_2 emissions	Virtually every modern building implements concrete to some extent	[11, 24–27]
Steel	Simple enough, different alloys allow for different types of steel	Less expensive than concrete, 6% cheaper	High tensile strength, different alloys offer different functionalities	More than concrete, 7–9% of global CO_2 emissions	Any large span building or skyscraper uses steel as a primary material	[11, 33–36]
Fired clay bricks	High temperature firing, energy-intensive	Low material cost, but energy intensive	Strong, durable, good thermal insulation	Can be recycled, but energy-intensive production	Easy to use, widely applicable in construction	[12]
Glass	High temperature melting, energy-intensive	High material and production cost	High transparency, poor insulation	100% recyclable, but energy-intensive	Used in windows, facades, and decorative items	[13]

[16]. In addition to their strength, concrete structures are also quite cheap, easy to mass produce and economical. This has led to concrete becoming the most used building material on Earth. An estimated 30 billion tonnes of concrete are used annually Worldwide. Per capita, this is over 3 times as much as 40 years ago. That means the demand for concrete materials is currently growing more steeply than that of wood or even steel [11]. As versatile and resilient as it may be, concrete has a colossal carbon footprint. The production and subsequent use of concrete alone accounts for around 8% of the total global CO_2 emissions, as shown in Fig. 6.1 [17–19].

This is a huge problem, since cement/concrete is actually expected to grow. In 2016, the global floor area was 235 billion square meters [20]. The Global Commission of Economy and Climate is estimating a total of $90 trillion to be invested in infrastructure alone by the year 2030, and that developing countries will account for two-thirds of this investment [21]. It is, therefore, imperative that the reduction of CO_2 emissions should start with the cement/concrete industry. This, however, is no simple task. On paper, there are plenty of methods and alternative materials that can help in reducing concrete CO_2 emissions [22]. It is not the lack of technology that may cause issues, it is the companies and countries which produce the cement. These producers need an incentive to transition, as well as a detailed outline on how to transition into a greener alternative. Fortunately, organizations such as the International Energy Agency (IEA) and the Global Cement and Concrete Association (GCCA) have both already published roadmaps on how to begin transitioning into using greener alternatives to concrete/cement [23, 24].

6.2.2 *Steel Construction*

The invention of steel as a reinforcement material has been revolutionary in architecture. Steel is a composite material, usually made from alloys of iron and carbon. Of course, different combinations of materials can create different alloys of steel all with their own unique properties [25]. In general, steel has both high strength and functionality. It is easy to work with and does not easily deform unless subjected to extremely high temperature or tremendous weight [26]. Steel also boasts a high strength to weight ratio while also being less time consuming to install when compared to concrete, it is also far more malleable and flexible. This makes steel an excellent choice in creating structural frameworks for towering modern buildings, as well as roofing, internal fixtures, and utilities [27]. In addition to everything mentioned so far, steel is also cheaper to ship than most building materials. Total building costs for steel structure was around 6% lower than concrete. For frame and upper floor costs, steel materials cost up to 18% less than concrete. What's more, steel structures take up to 7–10% less time to construct than concrete. In addition to this, steel structures also produced 10–20% less embodied carbon than concrete [28].

Steel production is highly energy-intensive and generates significant CO_2 emissions, therefore the energy required for steel production makes it a less sustainable option for zero-carbon buildings. While steel may be recyclable and a greener material than concrete, it is not an sustainable or zero carbon material. The steel industry contributes to between 7–9% of global CO_2 emissions, more than even concrete [29]. Figure 2 shows that steel production emits up to 2.6 gigatons of CO_2 $GtCO_2$ annually. According to the IEA, the steel sector is also the largest industrial consumer of coal, with it providing over 75% of the industries energy. Steel production remains a CO_2 intensive activity, one which is not showing any signs of slowing down. Recyclability of steel is a key feature of sustainability, but the production process remains unsustainable without advances in carbon–neutral methods.

In 2015, there was a noticeable step forward in regard to climate change, when 190 countries adopted the Paris Agreement. By 2019 The UK announces that itself along with the EU would pledge to total carbon neutrality by the year 2050. However, countries such as the USA, China and India noticeably never made such a pledge. These three nations are principal emitters, not just in regard to steel production emissions but overall emissions [30]. It sets a dangerous precedent for steel production, if the largest producers are not willing to commit to carbon neutrality.

6.2.3 Fired Clay Bricks and Glass

Traditional fired clay bricks are an environmentally intensive material due to the high temperatures required for firing, which consumes substantial energy (usually from fossil fuels) and significantly contributes to the carbon footprint. The global use of fired clay bricks is approximately 2.18 gigatons annually, resulting in approximately 1% of current global greenhouse gas emissions [12]. Coal is mostly used fuel for fired clay bricks contributing to high greenhouse gas emissions per produced brick [12].

Glass is essential material to provide natural daylight. The production of glass is energy-intensive and results in significant global CO_2 emissions, contributing around 86 million tons of CO_2 or approximately 0.3% of worldwide emissions. The manufacturing process involves melting raw materials at high temperatures, leading to substantial energy consumption. Additionally, transportation of glass products can further contribute to their carbon footprint [13].

6.3 Potentially Sustainable Materials

These materials are not necessarily sustainable due to the energy-intensive production process, but they have the potential to reduce the use of raw materials in construction, although recycling or other means can be considered a potentially sustainable feature. These materials have the potential to become more environmentally friendly with improvements in production processes, alternative sourcing, or technological advancements. Therefore, these materials are worth exploring for their future applications in ZEBs. Table 6.2 presents a literature summary on performances and sustainability features for potentially sustainable materials.

PET is not necessarily sustainable material since it is obtained from petroleum, and it is not biodegradable. Potentially sustainability feature comes from its recyclability, which decreases the need for raw plastic. Therefore, PET bricks are based on the PET recycling plastic waste, which may potentially keep waste out of landfills, reduce use of energy and carbon emissions, and minimize the use of natural resources, making them an eco-friendly alternative to conventional bricks. They are suitable for applications such as non-structural walls and other construction elements where lightweight materials are needed [31, 32].

6.3.1 Green Concrete

Green concrete may be a more sustainable alternative to traditional concrete. It often incorporates recycled materials like fly ash, slag, or silica fume to reduce the carbon footprint associated with cement production. Green concrete is not entirely zero-carbon, but it can be considered more sustainable than conventional concrete due to lower CO_2 emissions during production.

Green concrete implements cements that have a high content of one or more complementary cementing materials (CCM), such as reactive rice-husk ash, granulated slag, coal fly ash, and silica fume. It is worth noting that during the cement manufacturing process itself, a significant amount of CO_2 is directly released. However, carbon emissions are reduced when the amount of clinker is reduced by substituting with CCM. Research and development (R&D) into the use of CCM in Portland cements is still ongoing, with the possibility of up to 40% substitution by CCM [33]. In addition to CCM, there are also techniques and materials which focus on carbon capture and storage (CCS). These techniques are still somewhat new, though results on CO_2 reduction seem promising. Therefore, sustainable features of green concrete are waste reduction, use of industrial byproducts, lower carbon footprint than traditional concrete.

Green concrete can be defined as a concrete which incorporates waste material as one of its components, or its production process does not lead to environmental destruction [34]. Essentially, this is an environmentally friendly concrete, which adheres to the core principles of sustainability of having positive environmental, economic, and social impacts. There are a few key factors which determine if the concrete is green or not, such as the extent to which Portland cement is substituted with other materials, the way in which it is produced, its performance as well as factors that affect its durability. The main objectives of green concrete are to reduce CO_2 emissions from the cement industry, reduce the use of natural resources (limestone, shale, clay, etc..), as well as the inclusion of waste materials in concrete which cause pollution of natural resources (i.e. air, soil and water). Meeting these objectives would make green concrete viable for sustainable development without losing natural resources. Improving the cement used for concrete would also be important in solving

Table 6.2 Material performance and sustainability overview for sustainable materials

Material	Production	Cost	Functionality	Sustainability	Implementation	Ref
Green concrete	Simple as adding fly ash in to the cement mix	On par with conventional concrete	Generally, functions as well as conventional concrete	Sizeable reduction in CO_2 during concrete production stage	Used for retrofitting existing structures. However, there are issues with regulations	[48–51]
Carbon nanotubes (CNT)	Extremely difficult to manufacture	Expensive, exceeds even gold	Incredibly strong and durable at a tiny scale in small amounts	Extends life service of concrete thereby reducing lifetime CO_2 emissions of buildings	Requires significant strides in technology and research in order to start implementing as a construction material	[71, 71, 74, 75]
PET brick	• Heated to 200 °C, plastic acts as binder, sand as filler, or • Mixed using epoxy resin and clay, PET particles	Low to moderate	High compressive strength (up to 78.65 MPa), smooth surface, crack-free, ductile	Recycles plastic, environmentally friendly, low emissions	Can replace conventional bricks in non-structural walls	[31, 32]

issues in regard to longevity, upkeep and the environment. However, it might also be possible to entirely substitute cement with an alternative material. A case in point is the fly ash—a residue of coal combustion which is well known to be a viable cement substitute. For structural purposes, we would require only around 15–20% substitution. Fly ash is also readily available as a waste product of various coal-based power plants. Its utilisation has been proven in the mid-eighties of the last century when Canadian CANMET went on to produce a concrete mixture that consisted primarily of the fly ash. The result was a concrete with comparable mechanical properties, permeability, and good durability [35]. Higher quantities of fly ash (up to 50% by mass) used in the concrete mix have also shown promising results. This type of mix is known as HVFA (High Volume Fly Ash) concrete, and it has been proven to function in retrofitting concrete structures for buildings such as the Baker Hall building at the University of California, Berkley, USA [36]. Additionally, fly ash with a low content of calcium in combination granulated blast furnace slag produce a material known as Natural pozzolan. This material can also be used to substitute cement (again 50% of the mass) and create HVNP (High Volume Natural Pozzolan) which has been proven to achieve 14 MPa strength when aged 3 days, and a whole 38 MPa aged 28 days. Using fly ash in this case allows for reduction in CO_2 emissions during the cement manufacturing stage [37]. In 2000, UC Berkeley tested the resistance of HVFA concrete in case of the buildings within their native campus. The main idea was to provide a safe work/study in earthquake resistant buildings. Results found that the HVFA mixture not only reduced environmental impacts, but it also saved money while producing structures just as durable as regular concrete. This success has led the constructors of Barker Hall as well as Wurster Hall (the home to the College of Environmental Design) to 50% utilisation of HVFA in laying the foundations of these buildings. The production of all this required far fewer fossil fuels than conventional concrete [38].

6.3.2 Carbon Nanotubes

While carbon nanotubes (CNT) have outstanding properties (such as strength and conductivity), their production process is still quite energy-intensive, and not all CNT production methods are sustainable. However, CNTs could potentially help create stronger, lighter materials that reduce overall building material consumption. Although the manufacturing process of CNTs is still not sustainable due to substantial production emissions, the potential to reduce the use of materials in construction can be considered a sustainable feature of CNTs.

Due to its outstanding properties CNTs have received significant theoretical and experimental focus. CNT is quite similar in structure to activated carbons. Activated carbons are, without a doubt, an enormously effective and attractive material for capturing CO_2. AC are also arguably the most the most thoroughly researched out of any other carbon capture material [39]. As stated, CNT's have a similar structure to

AC's. Some of CNT's distinguishing features are its positive surface qualities, topological hollow tube architectures, great mechanical strength, extremely low weight, thermal and chemical stability, and a good pore size distribution [40–42]. The focus of their internal structure is the interactions between graphene layers that are rolled up into a cylinder with a hollow interior. This type of geometry enables us to control both the gas distribution and the residence time. Gas pressure, temperature, diameter across the material, and spacing between graphene layers all affect how much CO_2 is adsorbed in carbon nanotubes [43].

Carbon nanotubes (CNT) were first discovered in Russia in the early 50 s. However little research or development was done on them, possibly due to a lack of practical applications for CNTs at the time. They were re-discovered in the early 90 s by a Japanese research lab as a byproduct of fullerene synthesis [44]. CNTs have a very unique structure, they are made up of rolled graphite sheets. These sheets can be rolled in different ways to form tubes with slightly differing properties. Based on this it is possible to classify a few different types of CNTs [45]. However, this paper will focus on 2: Single-walled nanotubes (SWCNTs) and multi-walled nanotubes (MWCNTs). As their names imply, SWCNTs possess only one wall in a tubular shape, while MWCNTs, on the other hand, have multiple sliding walls. The diameter of CNTs stands anywhere between 1 and 100 nm in diameter—the size 50,000 times smaller than that of the diameter of human hair. The surface area is approximately between 100,000 and 700,000 m^2/kg [46]. The diameters of SWCNTs range from only a few nanometres (1–10 nm), while MWCNTs range anywhere in between 5–100 nm. This is because their structure consists of many concentric cylinders held together by van der Waals forces [47]. Interestingly, the tube lengths are unrestricted which means that they can be up to several micrometres or even millimetres. CNTs are one of the more promising materials of the twenty-first century. Their characteristics indicate the potential to significantly change and improve structural integrity [48], materials based on cement [49] as well as other fields [49–53]. CNTs are remarkably flexible (radial flexibility) they can sustain no damages even after been bent multiple times up to 90 degrees. They also have unparalleled compressive and tensile strength. Various types of nanotubes may exhibit tensile strengths above 60 GPa, with a theoretical tensile strength limit being around 200 GPa [54]. CNTs are no doubt a promising material with excellent mechanical, electrical, thermal, and chemical stability. One of the ways researchers have explored the real-world application of CNTs in through CNT reinforced cement-based materials. Studies have shown that adding a small percentage of CNT (0.048% by weight) will significantly improve the performance of cement-based materials [55]. CNT reinforced cement-based materials have a few advantages over conventional cement. Those being excellent mechanical properties, high sensing sensitivity, good durability, and good compatibility with the matrix. CNTs have shown to be good in reducing permeability, especially porosity and water absorption. In this way, CNTs prolong a lifespan of the matrix which can be useful in reducing total CO_2 emissions from buildings. CNT has great potential in cement-based composites. CNT could also become a potential 3D printed concrete (3DPC) material. Using CNT in 3DPC can offer various benefits including reduction in setting time for 3DPC as well as improved production efficiency, reduction in drying shrinkage of 3DPC, improvements in printing quality, and improvements in the early strength of 3DPC [56–60].

6.4 Sustainable Materials

Sustainable materials are those that have a lower environmental impact from extraction and production to disposal or recycling. These materials contribute to reduction building ecological footprint, promoting energy efficiency and reducing waste. Table 6.3 presents a literature summary on performances and sustainability features for selected sustainable materials.

6.4.1 Stabilized Mud Blocks

Masonry is one of the oldest building materials [63] with highly sustainable features. Stabilized Mud Blocks (SMB) as masonry are produced using minimal energy and local materials and can have a very low environmental impact. The use of locally sourced soil reduces transportation emissions, while the solar curing minimizes energy use and carbon emissions. Therefore, SMBs are considered as low-carbon and eco-friendly alternative to conventional fired bricks or concrete blocks. In essence, fired clay bricks are made by firing processed clay at a high temperature. Irreversible changes in clay minerals give bricks their strength, at the expense of a high energy cost. From an environmental perspective, energy conservation and clay mineral resources are significant in this case. Fired clay bricks can be replaced with sustainable, eco-friendly SMB. These are compacted solid blocks made from a soil, sand, stabilizer (lime), and water mixture. These blocks are used for wall construction after a 28-day curing period. The composition of the soil, the block's density, and the amount of stabilizer (lime or cement) all have a significant impact on the compressive strength of the block [64, 65].

Human civilizations have been using soil as a building material since ancient Mesopotamia and Egypt. However, mud by itself is not exactly the most optimal building material. It loses durability when exposed to rain and increases chances of termite infestations, which has led to mud being associated as the poor man's building material. This need not be the case; however, with only a few alterations mud has the potential to become a material that rivals and, in some aspects, exceeds conventional brick or concrete [66]. Made primarily from soil and stabilized with additional aggregates, earth blocks are a perfectly viable building material. There are a few types of earth blocks: Compressed Earth Block (CEB), compressed stabilized earth block (CSEB), and stabilized earth block (SEB). Stabilization as a technique simply means improving the properties of earth to grant it proper water absorption, durability, and stability (making sure its form, shape, size does not change under dry or moist conditions) all without firing [67, 68]. SMB are the result of combining and

Table 6.3 Material performance and sustainability overview for selected sustanable building materials

Material	Production	Cost	Functionality	Sustainability	Implementation	References
Stabilized mud blocks (SMB)	Can be done by small manual machines	Cheap, they cost less than bricks and concrete	Better than conventional adobe bricks, not as good as concrete	Less CO_2 generated during production, completely bio-degradable	Could be used as a substitute for bricks. However, they are not a popular option for developed nations	[54, 56–60]
Biochar	Slow with unsubstantial yields	Somewhat expensive due to the technology required to produce it	Fine-grain biochar helps to create stronger and more durable concrete	Raw versatility means it could possibly improve several areas of sustainability	Requires more research into mass production before any real implementation	[61, 63–65]
Wood	As easy as cutting a tree	Quite cheap	Excellent thermal insulation, 15 × better than brick and 400 × better than steel	Far more sustainable than any other conventional material	Mainly used for domestic homes, and smaller structures	[11, 43, 45–47]
PCMs	Made from organic/inorganic substances, encapsulated for stability or direct incorporation	High initial cost, but long-term energy savings	Stores/releases thermal energy to reduce interior temperature fluctuation	Reduces energy use, biodegradable, aids in achieving Zero Energy Buildings (ZEBs)	Used in walls, roofs, flooring and with active heating/cooling devices	[61, 62]

compressing sandy soil, a small amount of cement and lime at a specific moisture. The press used in this process only needs to be able to generate enough force to produce a dense block. Therefore, this stabilizer mix can be compressed using a manually operated press, reducing the need for expensive, complex and unsustainable machinery [69]. The blocks must be kept damp on their surfaces for three weeks in order for them to cure. Natural soil may be efficiently supplemented with additives such as fly ash, quarry dust from crushers, stone dust from stone cutting industries, and many other mining or industrial wastes [70]. This entire process consumes less energy (1/3 of the energy used for fired bricks) and thus produces far less CO_2 as a result. It is worth noting that different stabilizers in different amounts will result in a variety of earth blocks all with somewhat different properties. In general, SMB's offer a few advantages. Firstly, the compression strength of a properly made SMB's cast with a cement content of 7.5, 10 and 12% results in blocks ranging in strength between 7.03 MPa and 10.91 MPa. This means that, in terms of strength, SMB's can meet and even exceed typical adobe brick [71, 72]. SMBs are entirely biodegradable, easy to produce, and their production results in few CO_2 emissions making them an exceptionally sustainable option for developing nations. They have a low water absorption (mix of at least 10% cement have less than 15% water absorption) and they are also economical, with cost analysis of production showing that blocks with 10% cement are about 55.7% cheaper than fired bricks. Blocks having 10% cement 3% coir are about 17.14% cheaper than fired bricks [73]. SMBs provide a strong alternative to conventional bricks. They are inexpensive, easy to produce, reliable, sustainable, and we can already see them being used in countries such as Indonesia, India, Malaysia, etc. We could see this material being used extensively in developing nations as a substitute for even concrete.

6.4.2 *Biochar*

Biochar is a waste-derived substance with large-scale carbon sequestration capabilities. It is considered carbon–neutral or even carbon-negative depending on its production process. It sequesters carbon in the soil and can help improve soil quality in addition to being used in construction. However, the sustainability of biochar also depends on how it is produced (e.g., renewable biomass vs. fossil fuels). There is significant interest in the development of sustainable and low-carbon building materials improved with biochar. Because of its porous structure and highly functionalized surface, biochar is compatible with cement, asphalt, and polymer materials and can serve as a nucleation site for chemical processes. The most recent biochar-enhanced building materials, such as composites made of biochar and cement, plastic, and asphalt, were examined critically in this study. The functional qualities of biochar as construction materials were enhanced by elucidating its processes and efficacies [74, 75].

Biochar is a natural carbon-based material acquired through the process of pyrolysis. In short, the process involves heating biomass in a vacuum with no oxygen or oxidising agents, under temperatures ranging from 300–650 °C. Once the biomass is heated, three distinct products are created: gaseous (volatile parts), liquid (bio-oil), and solid (biochar) [76]. Biochar has a unique porous structure which allows for large storage capacity, all while being surprisingly cheap to produce. This means that it has outstanding potential as a CO_2 capture material [77]. Recently, biochar has been regarded as a potential substitute for cement and aggregates in building materials. Different types of char could fill different roles, for example fine-grain biochar can act as a cement substitute, while more coarse-grain char could possibly substitute aggregates in concrete [78]. A 2019 investigation on the yield stress and plastic viscosity of biochar-cement composites compared both coarse (2–100 μm) and fine (0.10–2 μm) biochar as a filler, where it was found that coarse biochar affected the viscosity of cement paste far more than fine biochar [79]. However, fine biochar particles were noted as stronger and had better water tightness early on. Fine biochar spreads into cement's pores and openings, thereby making it more compact and stronger [80]. It is worth noting that excessive biochar integration can and will weaken the stability and strength of biochar-cement. Therefore, it is imperative that biochar be subtly integrated in smaller amounts to ensure proper strength and stability. Nevertheless, even in small quantities integrated biochar can provide numerous benefits ranging from hydrothermal control and electromagnetic protection, to improved quality of air indoor, and better sound insulation [81–84]. The multi-purpose character of biochar puts it up as a material which has the potential to completely change how conventional construction materials are manufactured and thus become a future staple in sustainable development. With proper advances in research, development, technology and funding, we might possibly see the mass manufacture and extensive integration of building materials consisting of biochar.

6.4.3 Wood

Given how long it has been used, wood is possibly the oldest structural materials. Wood as a material has many advantages. It is inexpensive and readily available. Wood is a comparatively light building material, it is strong, long-lasting, and flexible. In addition to its impressive mechanical properties, wood is also an excellent thermal insulator [85]. The thermal insulation properties of wood are truly outstanding, thanks to the natural air pockets within its cellular structure. To put it into perspective in terms of thermal insulation, wood is 15 times better than masonry, 400 times better than steel, and 1770 times better than aluminium [86, 87]. As a result of this improved thermal performance, buildings produced using wood require less energy to heat and cool, which results in reduced energy bills. In terms of productivity, wood is one of the best materials. This is important since the UN has already estimated that, by 2030, the growing housing deficit will reach 240 million housing units. The question of how different nations deal with this deficit will be important for global sustainable development [88]. The housing sector is Wood is most often used in residential construction for houses, cottages, cabins, bungalows, etc.... Although, it is rare for buildings to be made entirely out of one material wood is often integrated into larger scale structures along with other materials. This leads into another attribute of wood, its ease of integration. Proper building construction ought to include multiple materials and methods to create the most stable and functional structures. Wood functions well with marble, aluminium, and steel all of which can be integrated into wood structures of additional structural stability.

Wood can be very sustainable if sourced responsibly from well-managed forests. It is a renewable resource, and trees absorb CO_2 as they grow, making it a carbon–neutral material when used in construction. The sustainability of wood depends on how it is harvested, processed, and transported. Only a few building materials possess the environmental benefits of wood. As with any other resource or material, it is crucial that the production phase releases as little CO_2 as possible in order to ensure sustainability. One of the wood's greatest attributes is the fact that it is a completely renewable and recyclable resource. If we make sure to follow proper sustainable forestry practices, we would never run out of wood. As a resource, it could potentially be endless and forever renewable [89, 90]. Another sustainable attribute of wood is its ability to absorb carbon and generates far lower greenhouse gas emissions than other materials during its production. For example, producing a ton of wood generates 33 kilos of net emissions, compared to 264 kilos for cement and 694 kilos for steel [91].

6.4.4 Phase Change Materials

Phase Change Materials (PCMs) can improve the energy efficiency of buildings by storing and releasing thermal energy, which can help reduce heating and cooling energy demand. PCMs be considered sustainable if produced from renewable sources, although overall sustainability depends on the material used and its lifecycle.

Depending on their specific characteristics and how they are produced. It is worth noting that not all PCMs may meet the criteria to be considered as sustainable but those whose attributes align with the goals of zero carbon construction, particularly in terms of improving energy efficiency [92] and reducing greenhouse gas emissions can be considered as sustainable. Therefore, to mitigate climate change PCMs can be an important component of sustainable building practices especially if combined with the other construction materials. To achieve net zero carbon emissions from building infrastructure PCMs work well if integrated in building structure or components [93].

Therefore, some PCMs can be considered sustainable if used to improve energy efficiency of buildings and reduce the need for heating and cooling (e.g. reduces reliance on mechanical heating/cooling) [94]. By decreasing energy consumption associated with temperature regulation, PCMs indirectly contribute to reducing carbon emissions during building operation.

6.5 Challenges and Limitations

Different sustainable materials and their possible uses as a new age of sustainable building technology are discussed in the literature. Green concrete is certainly an interesting prospect and has seen some real-world applications, it is still faced with the all-too-common issue of building regulations. Fly ash cement mixtures are still new and have not been fully tested or approved as building materials. Therefore, any buildings using green concrete would not pass building regulation. While buildings at UC Berkeley campus still use green concrete in their construction, these seem to be outliers. SMBs carry with them a fair few disadvantages including requiring proper soil identification, not being suitable for wide spans, high-rise and long buildings, generally low technical performance compared to concrete, and requiring quality checks at every stage of manufacturing to avoid low-quality products. In addition to all of that, there is the issue of optics. Most well-developed countries view materials such as SMBs as a staple of 3rd world developing countries, therefore it is unlikely that we will see many developed nations adopting SMBs as a construction material.

Biochar, while promising as a cement aggregate substitute, has severe issues in production. Production is quite a slow process, and the yields are not substantial, thus mass-produced biochar is still only theoretical. Major advances in technology and science and industry would have to be made to see a large-scale biochar integration into construction sector. CNTs are, by far, the most limiting ZCCM. Not only are they difficult to manufacture and currently cannot be mass produced for construction purposes, their cost of production and manufacture puts them as one of the most expensive materials on Earth surpassing even gold. There is also the issue of workability. If too much CNT is added to cement, then it can negatively affect workability. CNTs are also potentially a health hazard, with some instances of respiratory issues in animals as well as humans being directly linked to CNTs. However, this is still being researched and in general CNTs pose a lesser health risk than materials such as asbestos. Sustainable construction materials require further research and development before we might see a large-scale use for them. Table 6.4 summarizes the advantages and disadvantages.

However, current trends and concerns regarding climate change and CO_2 production might call for an increased effort on research in this area. It is not unlikely that we may see materials such as green concrete and SMBs used worldwide as a new standard within the next few decades. For biochar and CNT however, it is difficult to say whether we will see them fully integrated and realized as materials in construction within our lifetimes. Ultimately, whatever the future holds there is no doubt that sustainable construction materials are the next step in sustainability and healing the planet from the excessive pollution. Challenges and future trend on sustainable materials are presented in Table 6.5.

Table 6.4 Advantages and disadvantages

Material	Advantages	Disadvantages	References
Green concrete	Easy to manufacture, inexpensive, notably reduces overall CO_2 emissions for concrete, as strong and durable as conventional concrete	New material therefore it has not been extensively tested, construction regulations still do not recognize fly-ash cement as a proper building material	[48–52]
Stabilized mud bricks (SMB)	Simple to manufacture, very cheap, generally perform better than conventional adobe bricks, generate less CO_2 during production than bricks or concrete	Needs proper soil identification, not suitable for high-rise, low performance compared to concrete, requires quality checks at every stage of manufacturing	[53–60]
Biochar	Biochar can be integrated into cement to produce stronger and more durable, long-lasting sustainable concrete at a relatively cheap price. In addition to this, its raw versatility allows it to be used in various fields	Production of the biochar itself is a slow, expensive and difficult process. More research into the production and use of biochar is required, as it is still a fairly new and not well known material	[61–69]
Carbon nanotubes (CNT)	The strength of CNTs are unrivaled, small amounts of CNT added to concrete can extend its life service, which helps with overall CO_2 mitigation	Extremely difficult to manufacture, requires complex and specialized machinery, expensive to manufacture, costs more than even gold, health concerns in regards to toxicity within cases of long-term exposure	[70–86]
PCMs	Efficient thermal energy storage, Energy savings in heating/cooling, Temperature regulation, Can be sustainable	High initial cost, Limited temperature range, Durability issues, Complex encapsulation required	[95–97]

6.6 Concluding Remarks

The exploration of materials with sustainability features presents a significant opportunity in addressing the challenges of environmental degradation and the approaching energy crisis. The rising global CO_2 emissions, primarily driven by industry and fossil fuel combustion, have reached alarming levels, emphasizing the urgency for sustainable alternatives in construction. Traditional materials like concrete and steel, while critical to modern infrastructure, contribute significantly to global emissions, revealing a pressing need for environmentally friendly alternatives.

Table 6.5 Challenges and future trends

Material	Challenges	Future trends	Ref
Green concrete	Fly ash supply may lower as we move away from coal power. Noted issues with current building regulations	We might see green concrete as a viable alternative relatively soon, as long as the material meets current building regulations	[48–52]
Stabilized mud bricks (SMB)	SMB's main challenge is in fact the optics of the material itself. It is not a popular building material for more developed nations since it is perceived as a poor-, and low-quality material	It is possible that SMB's will become a staple in architecture for developing nations and mainly eastern and south-eastern countries. In fact, we may already see this happening in countries such as India, Indonesia, Malaysia, etc	[53–60]
Biochar	Biochar is a difficult material to process and produce, it takes expensive machinery and high heat in order to obtain biochar. This, combined with the fact that the yields are often quite low, represent the biggest challenge with integrating biochar as a construction material	It is difficult to say what the future holds for biochar. Significant efforts must be made in research and development before we see any large-scale biochar integration in any field. Although, its versatility would allow it to be used in virtually any field of human society	[61–69]
Carbon nanotubes (CNT)	The single greatest challenge in regard to CNT is its production. It is incredibly difficult, complicated, and expensive to produce. In addition to the abysmal yields, CNTs are far from ready to be mass manufactured or integrated as a construction material	Similar to biochar, CNTs require massive strides in research and technological development in order to reach mass production. Even in low amounts, they are very difficult to produce. As such, it is unlikely that we will see CNTs used as a building material in our lifetimes	[70–86]
PCMs	High cost of production and implementation Limited thermal conductivity reduces efficiency in some applications Degradation after multiple thermal cycles Encapsulation issues (leakage, material stability) Narrow temperature range for optimal performance	Integration with renewable energy systems (solar, wind) Smart PCMs that respond to dynamic environmental conditions Improved encapsulation methods for better stability and performance Cost reduction through new materials and mass production techniques Advanced applications in smart textiles, electronics, and energy-efficient buildings	[95–98]

Materials such as biobased materials, Green Concrete, Stabilized Mud Blocks, Biochar, and PCMs offer promising solutions. These materials not only show potential in reducing emissions but also complement the global shift towards sustainable development. However, the adoption of these materials faces several challenges, including technological limitations, regulatory obstacles, and economic constraints. Current construction practices and materials are deeply rooted, and transitioning to sustainable construction materials requires significant changes in manufacturing processes, building regulations, and market acceptance. Despite these challenges, the movement towards sustainable construction materials is inevitable given the growing awareness of climate change and the limited nature of fossil fuels. Future research and development will play a crucial role in shaping the construction industry, and their integration into mainstream construction practices could mark a turning point in our efforts to combat environmental degradation.

References

1. Friedlingstein P et al (2022) Global carbon budget 2022. Earth Syst Sci Data 14(11):4811–4900. https://doi.org/10.5194/essd-14-4811-2022
2. Kuo G (2019) When Fossil Fuels Run Out, What Then?—MAHB
3. Berardi U (2017) A cross-country comparison of the building energy consumptions and their trends. Resour Conserv Recycl 123:230–241. https://doi.org/10.1016/J.RESCONREC.2016.03.014
4. Nejat P, Jomehzadeh F, Taheri MM, Gohari M, Muhd MZ (2015) A global review of energy consumption, CO_2 emissions and policy in the residential sector (with an overview of the top ten CO_2 emitting countries). Renew Sustain Energy Rev 43:843–862. https://doi.org/10.1016/J.RSER.2014.11.066
5. Liu T, Tan Z, Xu C, Chen H, Li Z (2020) Study on deep reinforcement learning techniques for building energy consumption forecasting. Energy Build 208:109675. https://doi.org/10.1016/J.ENBUILD.2019.109675
6. Durakovic B, Torlak M (2017) Experimental and numerical study of a PCM window model as a thermal energy storage unit. Int J Low-Carbon Technol 12(3):272–280. https://doi.org/10.1093/ijlct/ctw024
7. Bernoville T What is the difference between carbon-neutral, net-zero and climate positive? plana.earth
8. Adelfio L, Giallanza A, La Scalia G, La Fata CM, Micale R (2023) Life cycle assessment of a new industrial process for sustainable construction materials. Ecol Indic 148:110042. https://doi.org/10.1016/J.ECOLIND.2023.110042
9. Luangcharoenrat C, Intrachooto S, Peansupap V, Sutthinarakorn W (2019) Factors influencing construction waste generation in building construction: Thailand's perspective. Sustainability 2019 11(13):3638. https://doi.org/10.3390/SU11133638
10. PCM-Based Building Envelope Systems—Innovative Energy Solutions for Passive Design | Benjamin Duraković | Springer. Accessed 15 Aug 2020 [Online]. Available: https://www.springer.com/gp/book/9783030383343
11. Monteiro PJM, Miller SA, Horvath A (2017) Towards sustainable concrete. Nature Publishing Group. https://doi.org/10.1038/nmat4930
12. Olsson JA, Hafez H, Miller SA, Scrivener KL (2025) Greenhouse gas emissions and decarbonization potential of global fired clay brick production. Environ Sci Technol 59. https://doi.org/10.1021/ACS.EST.4C08994/ASSET/IMAGES/LARGE/ES4C08994_0006.JPEG

13. Westbroek CD, Bitting J, Craglia M, Azevedo JMC, Cullen JM (2021) Global material flow analysis of glass: from raw materials to end of life. J Ind Ecol 25(2):333–343. https://doi.org/10.1111/JIEC.13112;SUBPAGE:STRING:FULL
14. Neville AM, Brooks JJ Concrete Technology—Neville & Brooks 2nd edition, 2010, Accessed 04 Dec 2023 [Online]. Available: https://toaz.info/doc-view-2
15. Mosa MJ, Jarallah HK, Salman MM (2025) Shear behavior of concrete deep beams with single coarse aggregate sizes. Herit Sustain Dev 7(2):1145–1162. https://doi.org/10.37868/HSD.V7I2.1663
16. Jackson MD et al (2014) Mechanical resilience and cementitious processes in Imperial Roman architectural mortar. Proc Natl Acad Sci USA 111(52):18484–18489. https://doi.org/10.1073/pnas.1417456111
17. Ellis LD, Badel AF, Chiang ML, Park RJY, Chiang YM (2020) Toward electrochemical synthesis of cement—An electrolyzer-based process for decarbonating $CaCO_3$ while producing useful gas streams. Proc Natl Acad Sci USA 117(23):12584–12591. https://doi.org/10.1073/pnas.1821673116
18. Olivier JGJ, Janssens-Maenhout G, Muntean M, Peters J (2016) Trends in Global CO_2 Emissions: 2016 Report; PBL Netherlands Environmental Assessment Agency: The Hague, PBL Netherlands Environmental Assessment Agency & European Commission's Joint Research Centre (JRC)
19. Ramsden K (2020) Cement and concrete: the environmental impact
20. Abergel T, Dean B, Dulac J Towards a zero-emission, efficient, and resilient buildings and construction sector [Online]. Available: www.globalabc.org
21. M. Bui et al. Carbon capture and storage (CCS): the way forward 01 May 2018. Royal Society of Chemistry. https://doi.org/10.1039/c7ee02342a
22. Kokulu N, Özgünler SA, Edis FE, Öztaş SK (2024) An LCIA-based model proposal for the selection of building interior finishing materials. Herit Sustain Dev 6(1):379–394. https://doi.org/10.37868/HSD.V6I1.425
23. GCCA (2021) Concrete Future—The GCCA 2050 Cement and Concrete Industry Roadmap for Net Zero Concrete
24. IEA (2018) Technology roadmap low-carbon transition in the cement industry. International Energy Agency
25. The World Steel Association (2013) The White Book of Steel
26. Khalafalla MA, Malyuta D (2013) review of different construction and building materials used before 21st century in Africa compared to the present situation after 21st century
27. Fathi MS (2010) Introduction to steel content schedule-5 th meeting
28. Theobald G "Steel Insight 3," 2012. Accessed 16 Dec 2023 [Online]. Available: chrome-extension://efaidnbmnnnibpcajpcglclefindmkaj/https://www.building.co.uk/download?ac=1724457
29. World Steel Association (2021) Climate change and the production of iron and steel
30. UN (2019) Report of the secretary-general on the 2019 climate action summit and the way forward in 2020
31. Adiyanto O, Ma'ruf F, Hopid A (2024) Assessing the optimal compressive strength of eco-friendly bricks using full factorial design. Sustain Eng Innov 6(1):119–130. https://doi.org/10.37868/SEI.V6I1.ID286
32. Ashraf N, El-Monayeri OD, Hassan HA (2024) Lego-like bricks manufacturing using recycled polyethylene (PE) and polyethylene terephthalate (PET) waste in Egypt. Sustainability 2024 16(19):8567. https://doi.org/10.3390/SU16198567
33. Oner A, Akyuz S, Yildiz R (2005) An experimental study on strength development of concrete containing fly ash and optimum usage of fly ash in concrete. Cem Concr Res 35(6):1165–1171. https://doi.org/10.1016/J.CEMCONRES.2004.09.031
34. Obla KH (2009) What is green concrete?
35. Bilodeau A, Mohan Malhotra V (2000) High-volume fly ash system: concrete solution for sustainable development. ACI Struct J 97(1). https://doi.org/10.14359/804

36. Manmohan D, Mehta PK (2002) Heavily reinforced shearwalls and mass foundations built with 'Green' Concrete. Concr Int 24(8)
37. Uzal B, Turanli L, Mehta PK (2007) High-volume natural pozzolan concrete for structural applications. ACI Mater J 104(5). https://doi.org/10.14359/18910
38. Schwarzenegger A (2005) Optimization of product life cycles to reduce greenhouse gas emissions in California. California Energy Commission
39. Li D et al (2020) Effects of compacting activated carbons on their volumetric CO_2 adsorption performance. Fuel 262:116540. https://doi.org/10.1016/J.FUEL.2019.116540
40. Ye Q, Jiang J, Wang C, Liu Y, Pan H, Shi Y (2012) Adsorption of low-concentration carbon dioxide on amine-modified carbon nanotubes at ambient temperature. Energy Fuels 26(4):2497–2504. https://doi.org/10.1021/ef201699w
41. Smart SK, Cassady AI, Lu GQ, Martin DJ (2006) The biocompatibility of carbon nanotubes. Carbon N Y 44(6):1034–1047. https://doi.org/10.1016/J.CARBON.2005.10.011
42. Rahimi K, Riahi S, Abbasi M, Fakhroueian Z (2019) Modification of multi-walled carbon nanotubes by 1,3-diaminopropane to increase CO_2 adsorption capacity. J Environ Manage 242:81–89. https://doi.org/10.1016/J.JENVMAN.2019.04.036
43. Belin T, Epron F (2005) Characterization methods of carbon nanotubes: a review. Mater Sci Eng, B 119(2):105–118. https://doi.org/10.1016/J.MSEB.2005.02.046
44. Iijima S (1991) Helical microtubules of graphitic carbon. Nature 354(6348):56–58. https://doi.org/10.1038/354056a0
45. Esawi AMK, Farag MM (2007) Carbon nanotube reinforced composites: potential and current challenges. Mater Des 28(9):2394–2401. https://doi.org/10.1016/J.MATDES.2006.09.022
46. Peigney A, Laurent C, Flahaut E, Bacsa RR, Rousset A (2001) Specific surface area of carbon nanotubes and bundles of carbon nanotubes. Carbon N Y 39(4):507–514. https://doi.org/10.1016/S0008-6223(00)00155-X
47. He XQ, Kitipornchai S, Wang CM, Liew KM (2005) Modeling of van der Waals force for infinitesimal deformation of multi-walled carbon nanotubes treated as cylindrical shells. Int J Solids Struct 42(23):6032–6047. https://doi.org/10.1016/J.IJSOLSTR.2005.03.045
48. Hu Y, Luo D, Li P, Li Q, Sun G (2014) Fracture toughness enhancement of cement paste with multi-walled carbon nanotubes. Constr Build Mater 70:332–338. https://doi.org/10.1016/J.CONBUILDMAT.2014.07.077
49. Ramezani M (2019) Design and predicting performance of carbon nanotube reinforced cementitious materials: mechanical properties and dipersion characteristics. University of Louisville. https://doi.org/10.18297/etd/3255
50. Amin MS, El-Gamal SMA, Hashem FS (2015) Fire resistance and mechanical properties of carbon nanotubes—clay bricks wastes (Homra) composites cement. Constr Build Mater 98:237–249. https://doi.org/10.1016/J.CONBUILDMAT.2015.08.074
51. Malarkey EB, Parpura V (2007) Applications of carbon nanotubes in neurobiology. https://doi.org/10.1159/000101885
52. Bandaru PR (2007) Electrical properties and applications of carbon nanotube structures. https://doi.org/10.1166/jnn.2007.307
53. Harrison BS, Atala A (2007) Carbon nanotube applications for tissue engineering. https://doi.org/10.1016/j.biomaterials.2006.07.044
54. Yakobson BI, Smalley RE (1997) Fullerene nanotubes: C 1,000,000 and beyond: some unusual new molecules—long, hollow fibers with tantalizing electronic and mechanical properties—have joined diamonds and graphite in the carbon family. Am Sci 85(4)
55. Sobolev K (2015) Nanotechnology and nanoengineering of construction materials. Nanotechnol Constr 3–13. https://doi.org/10.1007/978-3-319-17088-6_1
56. Chen Y, Figueiredo SC, Yalçinkaya Ç, Çopuroğlu O, Veer F, Schlangen E (2019) The effect of viscosity-modifying admixture on the extrudability of limestone and calcined clay-based cementitious material for extrusion-based 3D concrete printing. Materials 2019 12(9):1374. https://doi.org/10.3390/MA12091374
57. Mendoza Reales OA, Duda P, Silva ECCM, Paiva MDM, Filho RDT (2019) Nanosilica particles as structural buildup agents for 3D printing with Portland cement pastes. Constr Build Mater 219:91–100. https://doi.org/10.1016/J.CONBUILDMAT.2019.05.174

58. Chen J, Akono AT (2020) Influence of multi-walled carbon nanotubes on the hydration products of ordinary Portland cement paste. Cem Concr Res 137:106197. https://doi.org/10.1016/J.CEMCONRES.2020.106197
59. Chaipanich A, Nochaiya T, Wongkeo W, Torkittikul P (2010) Compressive strength and microstructure of carbon nanotubes–fly ash cement composites. Mater Sci Eng, A 527(4–5):1063–1067. https://doi.org/10.1016/J.MSEA.2009.09.039
60. Keyn M et al (2015) Influence of carbon nanotubes on the structure formation of cement matrix. IOP Conf Ser Mater Sci Eng 96(1):012046. https://doi.org/10.1088/1757-899X/96/1/012046
61. Duraković B (2020) Phase change materials for building envelope. In: PCM-based building envelope systems: innovative energy solutions for passive design, Springer Nature, pp 17–37. https://doi.org/10.1007/978-3-030-38335-0_2
62. Hasan A, McCormack SJ, Huang MJ, Norton B (2014) Energy and cost saving of a photovoltaic-phase change materials (PV-PCM) system through temperature regulation and performance enhancement of photovoltaics. Energies 7(3):1318–1331. https://doi.org/10.3390/EN7031318
63. Wani FM et al (2023) Finite element analysis of unreinforced masonry walls with different bond patterns. Sustain Eng Innov 5(1):58–72. https://doi.org/10.37868/SEI.V5I1.ID194
64. Minke G Introduction 1 Building with Earth Appendices 3 Building with Earth Design and Technology of a Sustainable Architecture Birkhäuser-Publishers for Architecture Basel · Berlin · Boston
65. Wiehle P, Brinkmann M (2022) Material behaviour of unstabilised earth block masonry and its components under compression at varying relative humidity. Case Stud Constr Mater 17:e01663. https://doi.org/10.1016/J.CSCM.2022.E01663
66. Jagadish KS (2007) Building with Stabilized Mud. I.K. International Publishing House Pvt. Ltd
67. Dutta AK, Deka D, Hatibaruah A, Goswami D, Rai SS, Basumatary T (2018) Environmental sustainability: stabilized mud block. https://doi.org/10.1007/978-3-319-74494-0_5
68. German Appropriate Technology Exchange, CRAterre (1995) Compressed earth blocks. Vieweg
69. Sajina C, RsuhasiniMTech (2015) Compressed stabilized mud block as an alternate building material [Online]. Available: www.ijrdt.org
70. Khaled Abdulrahman Y (2009) Durability properties of stabilized earth blocks
71. Amalkar MS, Ali M, Jagadish KS (2021) Strength of stabilized mud block masonry. J Build Pathol Rehabil 6(1). https://doi.org/10.1007/s41024-020-00101-2
72. Binici H, Aksogan O, Shah T (2005) Investigation of fibre reinforced mud brick as a building material. Constr Build Mater 19(4):313–318. https://doi.org/10.1016/J.CONBUILDMAT.2004.07.013
73. Prakash V, Raj A, Aravind S, Mathew B, Sumith VR (2016) Studies on stabilized mud block as a construction material. Int J Innov Res Adv Eng 3(01)
74. Fawzy S, Osman AI, Yang H, Doran J, Rooney DW (2021) Industrial biochar systems for atmospheric carbon removal: a review. Springer Science and Business Media Deutschland GmbH. https://doi.org/10.1007/s10311-021-01210-1
75. IPCC (2022) Strengthening and implementing the global response. In: Global warming of 1.5 °C. Cambridge University Press, pp 313–444. https://doi.org/10.1017/9781009157940.006
76. Racek J, Sevcik J, Chorazy T, Kucerik J, Hlavinek P (2020) Biochar—recovery material from pyrolysis of sewage sludge: a review. Waste Biomass Valorization 11(7):3677–3709. https://doi.org/10.1007/S12649-019-00679-W/METRICS
77. Chen WY, Mattern DL, Okinedo E, Senter JC, Mattei AA, Redwine CW (2014) Photochemical and acoustic interactions of biochar with CO_2 and H_2O: applications in power generation and CO_2 capture. AIChE J 60(3):1054–1065. https://doi.org/10.1002/AIC.14347
78. Akinyemi BA, Adesina A (2020) Recent advancements in the use of biochar for cementitious applications: a review. J Build Eng 32:101705. https://doi.org/10.1016/J.JOBE.2020.101705
79. Gupta S, Kua HW (2019) Carbonaceous micro-filler for cement: effect of particle size and dosage of biochar on fresh and hardened properties of cement mortar. Sci Total Environ 662:952–962. https://doi.org/10.1016/J.SCITOTENV.2019.01.269

80. Gupta S, Kua HW, Pang SD (2018) Biochar-mortar composite: manufacturing, evaluation of physical properties and economic viability. Constr Build Mater 167:874–889. https://doi.org/10.1016/J.CONBUILDMAT.2018.02.104
81. Park JH, Kim YU, Jeon J, Yun BY, Kang Y, Kim S (2021) Analysis of biochar-mortar composite as a humidity control material to improve the building energy and hygrothermal performance. Sci Total Environ 775:145552. https://doi.org/10.1016/J.SCITOTENV.2021.145552
82. Rodier L, Bilba K, Onésippe C, Arsène MA (2019) Utilization of bio-chars from sugarcane bagasse pyrolysis in cement-based composites. Ind Crops Prod 141:111731. https://doi.org/10.1016/J.INDCROP.2019.111731
83. Zou W, Gao B, Ok YS, Dong L (2019) Integrated adsorption and photocatalytic degradation of volatile organic compounds (VOCs) using carbon-based nanocomposites: a critical review. Chemosphere 218:845–859. https://doi.org/10.1016/J.CHEMOSPHERE.2018.11.175
84. Natalio F, Corrales TP, Feldman Y, Lew B, Graber ER (2020) Sustainable lightweight biochar-based composites with electromagnetic shielding properties. ACS Omega 5(50):32490–32497. https://doi.org/10.1021/ACSOMEGA.0C04639/ASSET/IMAGES/LARGE/AO0C04639_0005.JPEG
85. Kitek Kuzman M, Grošelj P Wood research 57 (4): 2012 591-600 wood as a construction material: comparison of different construction types for residential building using the analytic hierarchy process
86. Ladislaus Döry (President of EPF) and JOINT CONFERENCE EESC/EC/EP (2012) The Woodworking and Furniture sectors: proposing guidelines for a European sustainable and competitive industry
87. Ministry of Culture of The Republic of Lithuania and Architects Association of Lithuania, "National Concert Hall 'Tautos Namai'—Report VL1397"
88. UN (2023) The Sustainable Development Goals Report 2023 Special Edition
89. Dangel U (2016) Turning point in timber construction: a new economy
90. Falk B (2009) Wood as a sustainable building material. For Prod J 59(9)
91. Mahasenan N, Smith S, Humphreys K (2003) The cement industry and global climate change: current and potential future cement industry CO_2 emissions. In: Greenhouse gas control technologies—6th international conference, pp 995–1000. https://doi.org/10.1016/B978-008044276-1/50157-4
92. Duraković B (2020) PCMs in building structure. In: PCM-based building envelope systems: innovative energy solutions for passive design. Springer Nature, pp 63–87. https://doi.org/10.1007/978-3-030-38335-0_4
93. Sharma R, Jang JG, Hu JW (2022) Phase-change materials in concrete: opportunities and challenges for sustainable construction and building materials. Materials 2022 15(1):335. https://doi.org/10.3390/MA15010335
94. Simões N, Manaia M, Simões I (2021) Energy performance of solar and Trombe walls in Mediterranean climates. Energy 234:121197. https://doi.org/10.1016/J.ENERGY.2021.121197
95. Duraković B (2024) Phase change material applications in construction and building materials. In: Phase change materials for thermal energy management and storage, pp 231–260. https://doi.org/10.1201/9781003331957-10
96. Anupam BR, Sahoo UC, Rath P (2020) Phase change materials for pavement applications: a review. Constr Build Mater 247:118553. https://doi.org/10.1016/J.CONBUILDMAT.2020.118553
97. Ali HM et al (2024) Advances in thermal energy storage: fundamentals and applications. Prog Energy Combust Sci 100:101109. https://doi.org/10.1016/J.PECS.2023.101109
98. Khan MM, Alkhedher M, Ramadan M, Ghazal M (2023) Hybrid PCM-based thermal management for lithium-ion batteries: trends and challenges. J Energy Storage 73:108775. https://doi.org/10.1016/J.EST.2023.108775

Chapter 7
PCM Enhanced Construction Materials for Energy Storage

Abstract The use of construction materials that have the ability to store high energy densities is a promising way to promote energy-efficient practices in buildings. Phase change materials integration within concrete, mortar, brick, plasterboard, glass and wood have been investigated in this chapter. Particularly, various techniques, such as direct integration, immersion, impregnation, encapsulation and shape-stabilization are promising ways to improve building thermal inertia and thermal performances. It was found that certain challenges are associated such as potential leakage of PCM, compatibility with the base material, ensuring uniform distribution of PCM, and maintaining mechanical properties of base material. Achieving a balance between improved thermal performance and structural integrity is a trade-off that is unique to each material and integration method. Successful implementation depends on a thorough understanding of these methods, their advantages and challenges, enabling the creation of building materials that balance energy efficiency with practical functionality.

Keywords Sustainability · Energy storage · Latent storage materials · Phase change materials · Thermal performance · Energy efficiency

7.1 Introduction

Over the last decade, phase change materials (PCMs) became a very popular research topic especially in the context of building application due to its high latent heat. Building sector as the single largest user of raw materials and energy, with the help of PCMs has a unique opportunity for the development of more thermally efficient, and environmentally friendly energy storage materials. The energy density of these materials is much higher than those of the more common types, allowing them to store and release a large amount of energy within a certain temperature range. In the past decade, a lot of progress has made in PCM research and development as a result of the growing interest in combating climate change and embracing alternative energy sources [1]. The principle of PCMs involves endothermic and exothermic processes.

B. Duraković, *Zero Energy Buildings*, Green Energy and Technology,
https://doi.org/10.1007/978-3-032-16027-0_7

When the temperature of PCM rises, it absorbs heat and changes the phase of the material from solid to liquid and vice versa releasing the heat [2].

The pioneering research on phase change materials was initiated by Maria Telkes and Eleanor Raymond in the 1940s [3], although it did not attract significant attention until the energy crisis of the late 1970s and early 1980s. During this crisis, PCMs were extensively studied for various applications, especially in the applications in solar heating systems [4], as well as in other domains [5].

Heat storage within the building structure and components is considered as a passive storage system. Traditionally the passive storage system is based on Sensible Heat Thermal Energy Storage (SHTES). In this case the thermal energy is directly stored within the building structure and due to natural heat transfer (convection, conduction, and radiation) is released to the living space as needed [6]. Common materials used for building constructions are at the same time storage materials such as concrete, rock, brick, glass, adobe, plaster, wood, steel, etc. Since these materials have very limited storage capacity, PCMs have been investigated in several research papers as an enhancement for the building structure and components. PCMs have been applied in concrete [7], brick [8], gypsum plasterboards [9, 10], plaster [11], glazing systems [12, 13] and façade [14].

SHTES is the oldest form of thermal energy storage that has been used for centuries in buildings, known as building thermal mass. The increased thermal mass increases thermal inertia of buildings and significantly reduces variance of temperature and improves thermal comfort [15]. The storage capacity depends on the building structure materials [16] and the wall composition. Installing the insulation layer on the building exterior will improve the building thermal storage capacity [17].

This chapter aims to introduce latent heat thermal energy storage systems examining different methods for the incorporation of PCMs in construction materials. It discusses latent heat storage principles and various integration techniques into construction materials revolutionizing building thermal efficiency and energy sustainability. Also, it identifies a research gap that demands further investigation into the economic feasibility and long-term implications of PCM application in construction materials.

7.2 Phase Change Theory

In practical applications, PCMs are utilized in the form of pure substances, eutectic mixtures, or non-eutectic/binary mixtures. The key distinction lies in the phase change temperature between eutectic and non-eutectic mixtures. Eutectic PCM mixtures, being a special case of binary PCM mixtures, exhibit phase changes at a constant and lowest temperature compared to other binary mixtures. Conversely, non-eutectic/binary PCM mixtures undergo phase changes within a specific temperature range. Energy storage can take various forms, such as sensible heat in the solid/liquid phases, latent heat, or chemical energy (see Fig. 7.1). The transition between the solid and liquid phases of the material is responsible for storing latent heat energy, commonly referred to as melting, while the transition from liquid to solid is called solidification. The temperature at which PCM melting initiates is known as the solidus temperature, and the temperature at which the liquid phase of the PCM starts to freeze is referred to as the liquidus temperature [18].

In Fig. 7.1, melting and cooling curves of pure/eutectic PCM are shown. Sensible heat is accumulated during the heating process until the phase transition temperature (T_{PC}) is reached. The latent heat is accumulated during the melting (heating process) it is released during the solidification (cooling process). As temperatures exceed the PCM's melting point, the material undergoes a phase transition from solid to liquid, absorbing additional heat as the chemical bonds within the substance break. Conversely, when temperatures fall below the PCM's freezing point, the material releases the stored energy and reverts to its solid state [19]. Consequently, the amount of energy stored as sensible heat can be mathematically expressed as follows:

$$Q = \int_{T_1}^{T_2} mC_p dT = mC_p(T_2 - T_1) \tag{7.1}$$

The amount of energy stored in phase change materials (PCMs) can be mathematically expressed as the sum of sensible heat in the solid and liquid phases, along with the latent heat during phase transition.

$$Q = \int_{T_1}^{T_{pc}} mC_p dT + ma_{pc}\Delta h_{pc} + \int_{T_{pc}}^{T_2} mC_p dT \tag{7.2}$$

$$Q = mC_p\left(T_{pc} - T_1\right) + ma_{pc}\Delta h_{pc} + mC_p\left(T_{pc} - T_1\right) \tag{7.3}$$

The given equation considers several parameters: m for mass in kg, C_p for specific heat capacity in J/kg K, T_1 and T_2 for initial and final temperatures of the PCM in solid and liquid states in K, T_{pc} for the phase transition temperature of the PCM in K, Δh_{pc} for the enthalpy of the phase change material in J/kg, and a_{pc} as the fraction of melted PCM.

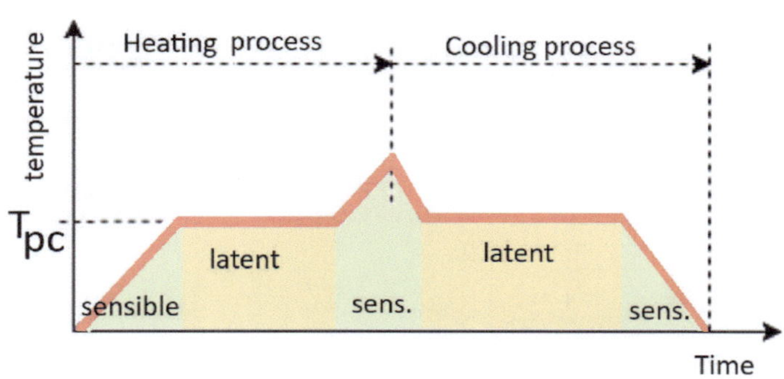

Fig. 7.1 Melting and solidification process. *Source* The Author

Equation (7.2) comprises three components: the first denotes sensible heat storage in the solid phase before reaching the phase change temperature (T_{PC}), the second represents latent heat of phase change, and the third signifies sensible heat storage in the solid phase after the completion of the melting process, with literature indicating that practical phase change materials for heat storage typically have phase change temperatures ranging from 20 to 60 °C [20].

7.3 Materials for Energy Storage

PCMs serve as thermal energy storage (TES) materials, utilizing latent heat during their phase change to collect or release energy. Particularly within a specific temperature range, PCMs can contribute to climate control. Different types and categories of PCMs exist including organic, inorganic, and eutectic PCMs [21]. Figure 7.2 shows the main PCMs commonly used in thermal energy storage applications.

Organic PCMs are commonly classified as paraffin and non-paraffin [22, 23]. These PCMs do not require supercooling during the freezing process [24], however, their major drawbacks are flammability [25], high production costs, and the use of nonrenewable resources like crude oil. They have very limited applications in building industry due to their high combustibility and safety concerns [26]. Paraffins are the most widely used in the studies due to their useful properties for the research. Non-paraffins including ethers, fatty acids, alcohols, and glycols, possess increased fire risk and reduced oxidation resistance [27–29].

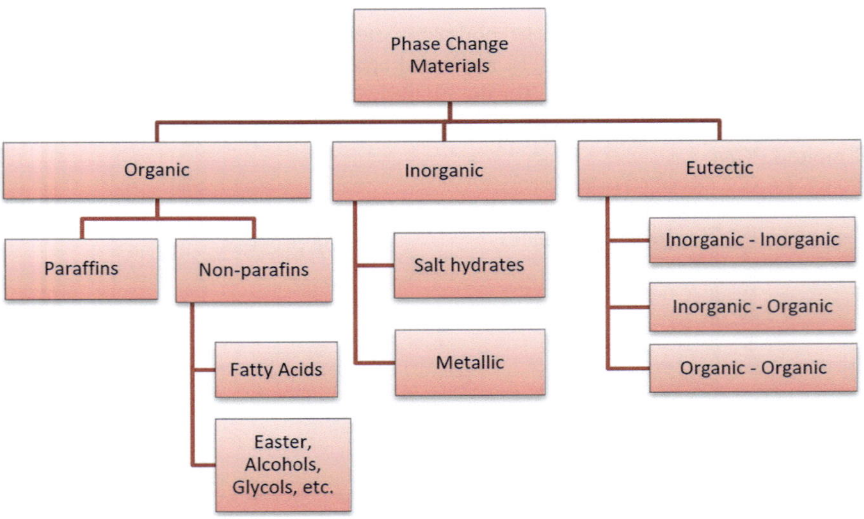

Fig. 7.2 Classification of phase change materials. *Source* The author

Inorganic PCMs are also commonly used in research (e.g. salt hydrates) due to their low production costs, high latent heat capacity and thermal conductivity [30]. A major drawback of salt hydrates is the potential for supercooling [31], phase segregation and corrosiveness, affecting their performance [32]. Metallics are suitable for high-temperature applications (ranging from 400 to 1000 °C) due to their excellent thermal and mechanical conductivity [33]. Typical applications of metallic PCMs are solar energy industry, reactors and furnaces where temperature control is required [34, 35]. Fatty acids PCMs are usually derived from plant or animal origin offering ecofriendly characteristics, low toxicity and cost, high latent heat capacity [36, 37]. As drawback is considered limited heat conductivity, which hinders their effectiveness as PCMs [37].

Eutectic PCMs are mixtures of two or more different phase change materials that melts/solidifies at sharp temperature. This temperature is lower than the melting points of the individual constituent components. Eutectics poses a high latent heat, good thermal cycling stability and less prone to phase separation compared to no eutectics. The drawback of eutectics is higher cost compared to the other PCMs [38]. Table 7.1 shows the major advantages and disadvantages of each PCM group.

Research conducted in recent years has led to the identification of numerous PCMs that are well-suited for latent heat storage applications. The extensive exploration of PCMs has yielded a diverse selection of materials with wide melting point

Table 7.1 Overview of advantages and disadvantages of PCMs

Categories of PCM materials	Advantages	Disadvantages	References
Organic PCMs	• No super cooling • No phase segregation • Low vapor pressure • Large temperature range • Self-nucleating • Compatible with conventional construction materials • Chemically stable • Recyclable	• Flammable • Low thermal conductivity • Low volumetric latent heat storage capacity	[30, 39]
Inorganic PCMs	• Low volumetric latent heat storage capacity • Higher thermal conductivity than organic PCMs • Low cost • Non-flammable • Sharp phase change	• Corrosive to metals • Supercooling • Phase segregation • Congruent melting • High volume change	[24]
Eutectics	• Sharp melting points • Properties can be tailored to match specific requirements	• Limited data on thermophysical properties for many combinations • High cost	[37]

Source The Author

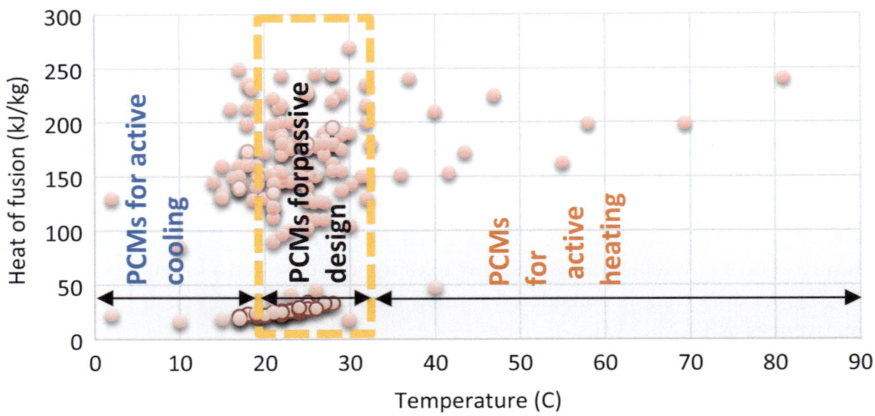

Fig. 7.3 PCMs for building applications. *Source* The Author

temperature ranges. This characteristic enables their application in a variety of conditions, making them highly adaptable for latent heat storage purposes. The latent heat stands as a critical property within PCMs, particularly in passive design applications, a higher latent heat is desirable. The heat of fusion and melting temperature is graphically depicted in Fig. 7.3, showing their relationship and as well as application range in active and passive designs.

PCMs suitable for passive building applications typically have specific melting temperatures and latent heat ranges in the range of 20–32 °C and up to 250 kJ/kg, respectively. These PCMs possess moderate to high thermal conductivities, making them efficient in transferring and storing thermal energy. Common examples of PCMs used in buildings include paraffin-based organic PCMs and salt hydrates, which offer a balance between cost-effectiveness, thermal performance, and ease of integration into building systems.

7.4 Incorporation Methods in the Construction Materials

PCMs can be seamlessly incorporated into building materials through a diverse array of methods including encapsulation, direct integration, absorption, and shape-stabilization techniques [40, 41]. Figure 7.4 shows four major methods of PCM incorporation in construction materials.

Encapsulation in macro and micro-capsules: encapsulated PCMs are capsules that contain the PCM. These capsules can be mixed into building materials. This approach prevents leakage, increases the surface area for better heat transfer, and avoids direct interaction between PCM and building materials [42]. In addition, the micro-encapsulated PCM may be used in construction goods in the form of additives [43], entire replacement, or sand replacement. Encapsulation processes

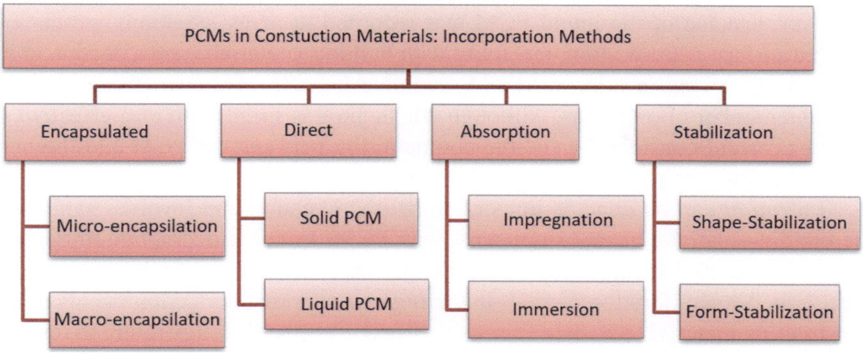

Fig. 7.4 Methods of incorporation of PCMs in construction materials. *Source* The Author

vary according to primary size, which affects shell structure and PCM choice. PCM encapsulated in shells with diameters more than 1 mm [44] or 1 cm [45] is termed as macro-encapsulation. The liquid PCM is injected directly into macro-encapsulated shells (relatively large containers such as tubes, polymer pouches, or others, usually larger than 1 cm in diameter). Since making shells smaller than 1 mm or 1 cm is difficult, micro-encapsulation requires more complicated methods. Spray drying [46] or interfacial polymerization [47] may be used to do this. For micro-encapsulation, PCM is usually spherical, while for macro-encapsulation, it might be rectangular, cylindrical, or spherical. Micro-encapsulation requires organic PCM, such as paraffin wax [48]. Figure 7.5 shows concepts of encapsulation of PCMs.

Direct Incorporation involves mixing of solid or liquid PCM directly into the building material during its manufacturing such as concrete, plaster, gypsum, and insulation materials (such as cellulose). This method is straightforward and relatively cost-effective. Direct inclusion of PCM and immersion in either solid or liquid phases, respectively, are the integration approaches that are the most cost effective [49, 50].

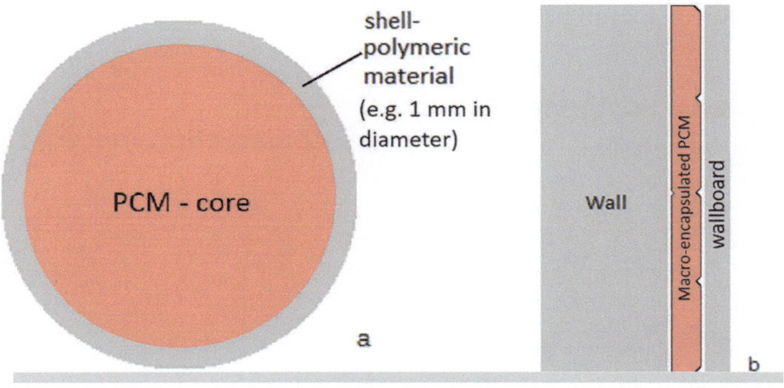

Fig. 7.5 Concept of encapsulations: **a** micro-encapsulation; **b** macro-encapsulation—flat plate container. *Source* The Author

However, leakage of the liquid PCM was recorded in many situations, producing esthetic surface discoloration, reactivity with building materials, and a decline in its performance owing to a loss of PCM's volume [51, 52]. This was found to have caused a number of problems. Although this method is simple and cost-effective, it requires careful consideration of the compatibility between the PCM and the host material to prevent segregation or performance degradation.

Absorption can be done as immersion or impregnation and involves saturating the building material or its components with PCM. In this process, building materials/components are dipped in liquid PCM, where it is absorbed in the material pore space by capillary forces [53]. In the *Impregnation* method, components of the construction material (e.g., aggregates) are dried at a specific temperature to remove moisture, and then soaked in molten PCM solution for a certain period. The pores of the aggregate are saturated with PCM, and the material's thermal properties have been enhanced. Another way is vacuum impregnation employs a pressure pump to eliminate entrapped air from the material pores. After this step, the vacuum chamber becomes a channel fishing of PCM into lightweight aggregate pores. In the last step, the air reenters the chamber to further push the PCM into the pores [54, 55].

Immersion is similar to impregnation, which involves incorporating PCM by soaking porous construction material (e.g., concrete product) in molten PCM. This approach is cost-effective but has certain challenges such as leakage and interaction between the liquid PCM and the building material. Leakage can result in aesthetic deterioration and diminished performance due to PCM loss, while interactions with the building material might trigger adverse chemical reactions. As such, while immersion presents economic advantages, careful consideration is necessary to mitigate these potential drawbacks and ensure the long-term effectiveness of the integrated PCM solution. The effectiveness of this method depends on factors such as the type of PCM, temperature during immersion, and absorption capacity of the concrete [56]. Figure 7.6 shows conceptual approaches for the absorption method.

Shape-Stabilization involves incorporating PCM into the building material in a way that maintains its shape. The incorporation of PCM is crafted by infusing a carrier material with PCM, a design that effectively prevents PCM leakage during repetitive cycles of melting and solidification [17]. This approach is useful when

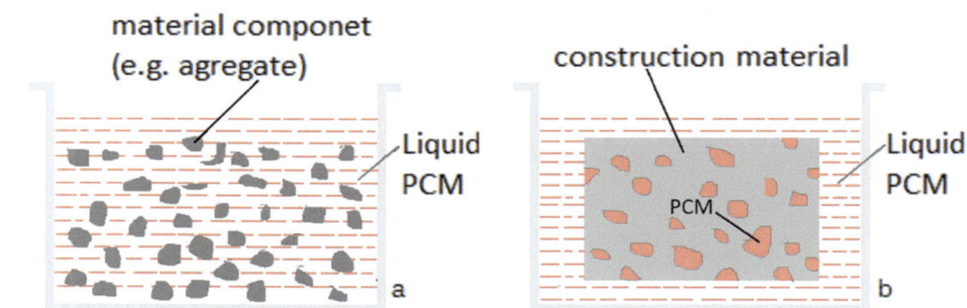

Fig. 7.6 Concept of absorption: **a** impregnation; **b** immersion. *Source* The Author

structural integrity is important. In the development of Shape-Stabilized Phase Change Materials (SSPCMs), PCM is integrated into a supporting matrix, which can include materials like expanded vermiculite [57], expanded graphite, expanded perlite [58], diatomite, high-density polyethylene, butadiene, carbon nanotubes, styrene, expanded perlite [58, 59], expanded recycled glass aggregate, recycled brick aggregate [60], zeolite [61], diatomite [62] and more. This type of composite has gained significant traction in the past decade. Notably, SSPCMs have garnered attention due to their ability to retain their shape throughout multiple heat cycles without the need for encapsulation [63–65]. Similarly, Form-Stabilized Composite Phase Change Materials (FSCPCM), are solid–liquid PCMs integrated into matrices with higher melting temperatures, which act as a sponge and prevent PCM free flow above the melting point [66]. This immobilization, often referred to as shape-stabilization, is especially relevant when the embedded PCM remains in a solid-state during phase transitions [67]. Figure 7.7 represents a conceptual approach to SSPCM integration. Figure 7.7 shows the conceptual approach to shape-stabilized composite PCMs.

Coating of the construction materials is an important aspect during the incorporation of PCMs in construction materials. To prevent any PCM leakage caused by thermal cycles, it is essential to apply suitable coating to the PCM-infused aggregates. The coating ensures the consistent performance of PCM and serves as a barrier preventing PCM leakage from impregnated or immersed construction materials. An epoxy resin can be used as a coating material [68]. In addition, coating enables the controlled release and absorption of heat from specific areas, contributing to localized temperature regulation.

On the other hand, coating can restrict the volume available for PCM storage hich may reduce the overall energy storage capacity of the system. During the coating and drying process, PCM may experience coalescing, leading to the formation of clusters or uneven distribution. Coalescing can impact the efficiency and uniformity of the thermal storage process. Over time and due to various factors, the coating might undergo delamination or detachment, compromising its protective function. This can lead to PCM leakage and diminish the intended thermal performance [68, 69]. Each integration method has its own set of advantages and challenges that are summarized in Table 7.2.

Choosing the appropriate integration method depends on factors such as the intended application, the host material, desired thermal performance, and overall

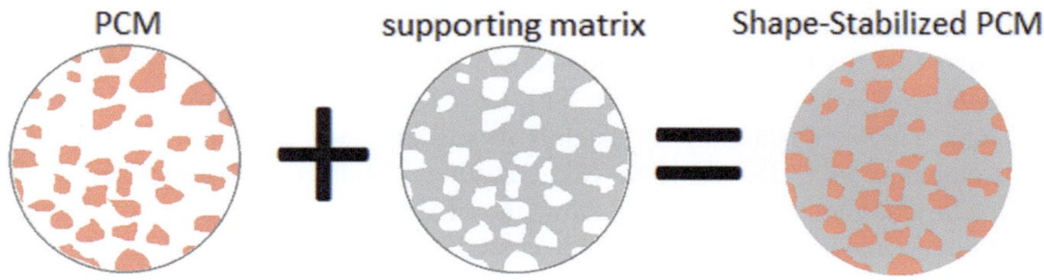

Fig. 7.7 Concept of shape-stabilized composite PCM. *Source* The Author

Table 7.2 Advantages and challenges of PCM integration methods

Method	Advantages	Challenges	Reference
Direct incorporation	Simplicity and cost-effectiveness	Compatibility and potential material degradation (e.g. interferences with cement hydration and mechanical properties reduction)	[70–72]
Encapsulation	Leak prevention, enhanced thermal conductivity, reduction in reaction of PCM with the surrounding, and improved durability	Increased material complexity and potential capsule breakage	[73]
Impregnation and immersion	Cost-effectiveness and ease of implementation	Leakage, reduced material strength, and potential interactions	[68]
Shape-stabilization	Increase heat transfer and prevent PCM leakage Maintained shape and structural integrity	Requires specialized processing and might affect material properties	[74]
Coating	Prevents leakage of the PCM from impregnated/immersed construction materials or from micro-encapsules Localized temperature control	Limited energy storage capacity, potential coating delamination Coalescing of the PCM during the coating and drying process	[68, 69]

Source The Author

project requirements. Proper selection of PCM and the integration method withing construction in using full PCM-enhanced construction materials potential.

7.5 PCM-Based Construction Materials

PCMs have found application in a variety of building materials, which represents a promising way to improve energy efficiency and optimize thermal performance in buildings [75]. Research shows that PCM can be applied into various construction materials, such as concrete pavements [76, 77] and asphalt deicing pavements [78, 79]. Also, they are integrated into the building structure itself, including bricks [80, 81], solar walls [13], concrete walls [82], masonry [83], and plasters and mortars [84, 85]. Their application also extends to ceilings [86, 87], plasterboard [88], glass elements [89–91], and roofs [87], creating a wide range of applications in the construction industry. Table 7.3 shows the advantages and challenges of selected construction materials that have been discussed in literature.

The application of PCMs in foam concrete holds the potential to improve thermal storage capacity and insulation within building structures [16, 93]. The utilization of

Table 7.3 PCM-based construction materials advantages and challenges

Material	Advantages	Challenges	Ref
PCM-based foam concrete	• Enhanced thermal storage capacity • Improved thermal performance • Lightweight and porous structure	• Precise mix proportions • Mechanical properties may be affected	[16, 75, 92–96]
Concrete	• Reduced thermal conductivity • Increased thermal mass • Lower interior temperatures in warm climates	• Lower strength • Long-term stability concerns	[82, 97]
Concrete pavement	• Efficient temperature management • Reduced freeze–thaw damage • Lower deicing needs	• High compressive strength requirements • Challenges during construction	[98–100]
PCM-based cement render	• • Improved thermal properties • Suitable for various construction applications	• Challenges related to workability and phase change behavior	[9, 101, 102]
Gypsum boards	• Decreased cooling loads • Enhanced thermal inertia • Potential for energy flexibility	• Impact on physical and mechanical properties • PCM retention and fire resistance	[9, 102, 103]
Wood	• Elevated energy storage and temperature regulation • Renewable and porous material	• PCM retention and fire safety concerns	[104, 105]
Cladding materials	• Improved thermal insulation • Reduced heat gain from external environment	• Positioning of PCM layer • Impact on thermal and solar transmittance	[106, 107]
Fillings in PCM brick	• Enhanced thermal resistance and inertia Improved thermal performance	• Balancing thermal resistance and inertia • PCM quantity and retention • Precise PCM quantity adjustments • Impact on mechanical properties	[108–110]

distinct phase transition temperatures tailored to specific climatic conditions is key, particularly when considering foam concrete formulated from cement, sand, and water [16, 93]. Incorporating a foaming agent further introduces sporadic air pockets into the mix, enhancing porosity [94]. Nonetheless, attaining precise mix proportions remains a challenge in realizing the desired attributes [111]. A compelling aspect of foam concrete is its inherent compatibility with PCM capsule embedding, providing a porous and lightweight framework without compromising structural integrity [95].

Paraffin PCM emerges as a promising choice for its cost-effectiveness and stability [112]. Within foam cement, a suitable composite is one with 45% paraffin content, keeping a balance between structural stability and thermal performance [96]. The phase-change foam concrete surpasses traditional foam concrete in terms of thermal conductivity and heat absorption capacity [96]. Furthermore, the thermal conductivity of composite PCM could potentially decrease as its composition increases, translating to heightened heat storage capacity [96].

Direct incorporation of PCM into concrete yields decreased thermal conductivity and increased thermal mass with trade-offs such as diminished mechanical strength and fire resistance [82]. Integration of PCM within drilled holes or voids showed potential benefits such as reduced interior temperatures in warm climates [97]. The choice of PCM, in this case, paraffin with a 27.5 °C melting point, led to decreased temperatures on the opposite side of the hollow concrete during summer [97]. It is important to validate these findings with real-world weather conditions [97].

Concrete pavements often suffer damage due to thermal dilatation caused by extreme temperatures. The concrete expands during the hot season and may crack when frozen. PCM incorporation in pavements (placing them within pipes, encapsulated particles, or surface voids) may mitigate temperature extremes and freeze–thaw damage [113]. On the other hand, temperature fluctuations and heavy loads require durable encapsulation and sufficient structural. Suitable PCMs for this application are those that have temperature transition near-zero Celsius [98].

The PCM-based cement render at 60% of PCM content appears viable but still there are challenges related to workability, mechanical properties and phase change behavior [114], showed its potential for diverse construction applications [99].

The utilization of PCMs in gypsum boards increases their thermal storage capacity and showed the potential for decreased cooling loads and improve thermal inertia of building [9]. Since PCM integration may affect mechanical properties, reinforcements can counteract these effects [101], offering a prospective material for energy-efficient construction [115]. The thickness and position of PCM-layer gypsum boards is important in achieving optimal thermal exchange and energy efficiency [102]. The optimum PCMs positions dependend on a daily melting/freezing cycle to be ready for the following day. In many cases it is close to the interior surface when the objective is to stabilize indoors or closer to the exterior surface when the objective is to reduce external heat gain [116]. PCM layers applied to the exterior surfaces of buildings have the potential to improve thermal insulation, especially when strategically placed according to local climatic conditions [106, 107]. Such layers can reduce heat transfer from the external environment, thereby reducing the energy required to operate cooling systems [106].

Table 7.4 PCM-based construction materials properties

Material	Latent heat (J/kg)	Average thermal conductivity (W/m K)	Average compressive strength (MPa)	Average density (kg/m^3)	References
Foamed concrete		0.53	32.38 (over 7 days)	896	[119–121]
Cement render	180	0.54	14.5 (over 28 days)	1392	[52]
Cement render	180	0.18	4.9	896	[122]
Brick	190 - 7360	0.21–0.358	8.73–23.75	930–2049	[123–125]
Gypsum board	1.13 (sensible)	0.15–0.417	2.5–4.8	956–1120	[126–128]
Glass	15.3–18.4	0.2–0.280	15–1000	750–880	[129, 130]
Wood	21.12–25.12	0.08–0.32	30–60	315–900	[131, 132]

Source Authors

The integration of PCM within wood-based materials improves its energy storage capacity and temperature regulation [104]. The infusion of PCM into wood composites capitalizes on the material's mechanical strength and inherent thermally induced flexibility [117]. Addressing challenges such as PCM retention and fire resistance are key to realizing the benefits of integrating PCM into wood-based materials [118]. The unique properties of wood in absorbing and releasing significant amounts of heat make wood composites with integrated PCM an important contribution to energy efficiency [105, 107].

Table 7.4 provides a brief overview of average values of the PCM-based construction material properties such as density, compressive strength, thermal conductivity, or latent/sensible heat.

The integration of PCMs into various construction materials provides a range of opportunities for building energy efficiency and optimizing thermal performance. Despite challenges involving material compatibility, structural integrity, and fire resistance, the potential benefits in terms of reduced energy consumption and improved thermal performance are still opportunities for building sector [133].

7.6 Conclusion

PCM incorporation methods that have been considered include direct integration, immersion, impregnation, encapsulation, and shape-stabilization, each of which offers specific advantages and challenges. The materials discussed in this chapter, such as foamed concrete, cement mortar, brick, plasterboard, glass and wood, possess various properties that make them suitable for PCM applications.

Despite certain benefits of the integration of PCMs in construction material, its widespread adoption faces certain challenges. The advantages of each method must be carefully considered against the associated challenges, such as maintaining material strength, preventing leakage, long-term stability, fire resistance, and compatibility issues. Ensuring the durability and longevity of PCM-based materials is crucial, as are addressing PCM leakage issues and integration methods. As research progresses, optimization of these methods and developing hybrid approaches may foster innovative solutions for PCM integration. The exploration of innovative encapsulation techniques, improved compatibility between PCMs and matrices, and improved manufacturing processes may lead to materials that effectively utilize the benefits of phase change for sustainable building design.

References

1. Phase change materials: applications in the built environment|NBS. Accessed 11 Dec 2022. [Online]. Available: https://www.thenbs.com/knowledge/phase-change-materials-applications-in-the-built-environment
2. Rathod MK, Kanzaria HV (2011) A methodological concept for phase change material selection based on multiple criteria decision analysis with and without fuzzy environment. Mater Des 32(6):3578–3585. https://doi.org/10.1016/J.MATDES.2011.02.040
3. Telkes M, Raymond E (1949) Storing solar heat in chemicals—a report on the Dover house. Heat Vent 46(11):80–86
4. Sokolov M, Keizman Y (1991) Performance indicators for solar pipes with phase change storage. Sol Energy 47(5):339–346. https://doi.org/10.1016/0038-092X(91)90027-T
5. Rathod MK, Banerjee J (2013) Thermal stability of phase change materials used in latent heat energy storage systems: a review. Renew Sustain Energy Rev 18:246–258. https://doi.org/10.1016/J.RSER.2012.10.022
6. Duraković B (2020) PCMs in separate heat storage modules. In: PCM-based building envelope systems: innovative energy solutions for passive design. Springer Nature, pp 121–146. https://doi.org/10.1007/978-3-030-38335-0_6
7. Thiele AM, Sant G, Pilon L (2015) Diurnal thermal analysis of microencapsulated PCM-concrete composite walls. Energy Convers Manag 93:215–227. https://doi.org/10.1016/j.enconman.2014.12.078
8. Vicente R, Silva T (2014) Brick masonry walls with PCM macrocapsules: an experimental approach. Appl Therm Eng 67(1–2):24–34. https://doi.org/10.1016/J.APPLTHERMALENG.2014.02.069
9. Oliver A (2012) Thermal characterization of gypsum boards with PCM included: thermal energy storage in buildings through latent heat. Energy Build 48:1–7. https://doi.org/10.1016/J.ENBUILD.2012.01.026
10. Karaipekli A, Sari A (2016) Development and thermal performance of pumice/organic PCM/gypsum composite plasters for thermal energy storage in buildings. Sol Energy Mater Sol Cells 149:19–28. https://doi.org/10.1016/J.SOLMAT.2015.12.034
11. Haurie L, Serrano S, Bosch M, Fernandez AI, Cabeza LF (2016) Single layer mortars with microencapsulated PCM: study of physical and thermal properties, and fire behaviour. Energy Build 111:393–400. https://doi.org/10.1016/J.ENBUILD.2015.11.028
12. Duraković B (2020) PCM-Based glazing systems and components. In: PCM-based building envelope systems: innovative energy solutions for passive design. Springer Nature, pp 89–119. https://doi.org/10.1007/978-3-030-38335-0_5

13. Durakovic B, Halilovic M (2023) Thermal performance analysis of PCM solar wall under variable natural conditions: an experimental study. Energy Sustain Dev 76:101274. https://doi.org/10.1016/J.ESD.2023.101274
14. de Gracia A, Navarro L, Castell A, Ruiz-Pardo Á, Alvárez S, Cabeza LF (2013) Experimental study of a ventilated facade with PCM during winter period. Energy Build 58:324–332. https://doi.org/10.1016/J.ENBUILD.2012.10.026
15. Pires L, Silva PD, Castro Gomes JP (2013) Experimental study of an innovative element for passive cooling of buildings. Sustain Energy Technol Assess 4:29–35. https://doi.org/10.1016/J.SETA.2013.08.002
16. Findik F (2022) Green concrete for structural buildings. Herit Sustain Dev 4(1):67–76. https://doi.org/10.37868/HSD.V4I1.84
17. Yildiz G, Duraković B, Abd Almisreb A (2021) Performances study of natural and conventional building insulation materials. Int J Adv Sci Eng Inf Technol 11(4):1395–1404. https://doi.org/10.18517/IJASEIT.11.4.11139
18. Duraković B (2020) Phase change materials for building envelope. In: PCM-based building envelope systems: innovative energy solutions for passive design. Springer Nature, pp 17–37. https://doi.org/10.1007/978-3-030-38335-0_2
19. Madad A, Mouhib T., Mouhsen A (2018) Phase change materials for building applications: a thorough review and new perspectives. Buildings 8(5):63. https://doi.org/10.3390/BUILDINGS8050063
20. Shukla A, Buddhi D, Sawhney RL (2009) Solar water heaters with phase change material thermal energy storage medium: a review. Renew Sustain Energy Rev 13(8):2119–2125. https://doi.org/10.1016/J.RSER.2009.01.024
21. Liu L, Hammami N, Trovalet L, Bigot D, Habas JP, Malet-Damour B (2022) Description of phase change materials (PCMs) used in buildings under various climates: a review. J Energy Storage 56:105760. https://doi.org/10.1016/J.EST.2022.105760
22. Chen Q, Wang H, Gao H, Wang X, Ma B (2022) Effects of porous silicon carbide supports prepared from pyrolyzed precursors on the thermal conductivity and energy storage properties of paraffin-based composite phase change materials. J Energy Storage 56:106046. https://doi.org/10.1016/J.EST.2022.106046
23. Ling TC, Poon CS (2013) Use of phase change materials for thermal energy storage in concrete: an overview. Constr Build Mater 46:55–62. https://doi.org/10.1016/J.CONBUILDMAT.2013.04.031
24. Casini M (2016) Phase-change materials. Smart Build 179–218. https://doi.org/10.1016/B978-0-08-100635-1.00005-8
25. Phase Change Materials The report provides a review of Phase Change Materials (PCMs) for Thermal
26. Paraffin as Phase Change Material|IntechOpen. Accessed 11 Dec 2022. [Online]. Available: https://www.intechopen.com/chapters/70480
27. Stritih U (2003) Heat transfer enhancement in latent heat thermal storage system for buildings. Energy Build 35(11):1097–1104. https://doi.org/10.1016/J.ENBUILD.2003.07.001
28. Aydn AA, Okutan H (2011) High-chain fatty acid esters of myristyl alcohol with odd carbon number: novel organic phase change materials for thermal energy storage—2. Sol Energy Mater Sol Cells 95(8):2417–2423. https://doi.org/10.1016/J.SOLMAT.2011.04.018
29. Tyagi VV, Kaushik SC, Tyagi SK, Akiyama Y (2011) Development of phase change materials based microencapsulated technology for buildings: a review. Renew Sustain Energy Rev 15(2):1373–1391. https://doi.org/10.1016/J.RSER.2010.10.006
30. Tebaldi ML, Belardi RM, Montoro SR (2016) Polymers with nano-encapsulated functional polymers: encapsulated phase change materials. encapsulated phase change materials. In: Design and applications of nanostructured polymer blends and nanocomposite systems, pp 155–169. https://doi.org/10.1016/B978-0-323-39408-6.00019-4
31. Ushak S, Judith Cruz M, Cabeza LF, Grágeda M (2016) Preparation and characterization of inorganic PCM microcapsules by fluidized bed method. Materials 9(1). https://doi.org/10.3390/MA9010024

32. Xie N, Huang Z, Luo Z, Gao X, Fang Y, Zhang Z (2017) Inorganic salt hydrate for thermal energy storage. Appl Sci 7(12):1317. https://doi.org/10.3390/APP7121317
33. Mohamed SA et al (2017) A review on current status and challenges of inorganic phase change materials for thermal energy storage systems. Renew Sustain Energy Rev 70:1072–1089. https://doi.org/10.1016/J.RSER.2016.12.012
34. Khare S, Dell'Amico M, Knight C, McGarry S (2012) Selection of materials for high temperature latent heat energy storage. Sol Energy Mater Sol Cells 107:20–27. https://doi.org/10.1016/J.SOLMAT.2012.07.020
35. Nomura T, Zhu C, Sheng N, Saito G, Akiyama T (2015) Microencapsulation of metal-based phase change material for high-temperature thermal energy storage. Sci Rep 5. https://doi.org/10.1038/SREP09117
36. Narayanan SS et al (2017) Development of sunlight-driven eutectic phase change material nanocomposite for applications in solar water heating. Resour Effic Technol 3(3):272–279. https://doi.org/10.1016/J.REFFIT.2016.12.004
37. Nazari M, Jebrane M, Terziev N (2020) Bio-based phase change materials incorporated in lignocellulose matrix for energy storage in buildings—a review. Energies 13(12):3065. https://doi.org/10.3390/EN13123065
38. Rathod MK, Rathod MK (2018) Thermal stability of phase change material. Phase Chang Mater Their Appl. https://doi.org/10.5772/INTECHOPEN.75923
39. Rathore PKS, Shukla SK (2021) Enhanced thermophysical properties of organic PCM through shape stabilization for thermal energy storage in buildings: a state of the art review. Energy Build 236:110799. https://doi.org/10.1016/J.ENBUILD.2021.110799
40. Aridi R, Yehya A (2022) Review on the sustainability of phase-change materials used in buildings. Energy Convers Manag: X 15:100237. https://doi.org/10.1016/J.ECMX.2022.100237
41. Duraković B (2024) Phase change material applications in construction and building materials. In: Change materials for thermal energy management and storage, pp 231–260. https://doi.org/10.1201/9781003331957-10
42. Editorial Board. J Energy Storage 24:100857. https://doi.org/10.1016/s2352-152x(19)30712-1
43. Theodoridou M, Kyriakou L, Ioannou I (2016) PCM-enhanced lime plasters for vernacular and contemporary architecture. Energy Procedia 97:539–545. https://doi.org/10.1016/J.EGYPRO.2016.10.070
44. Akeiber H et al (2016) A review on phase change material (PCM) for sustainable passive cooling in building envelopes. Renew Sustain Energy Rev 60:1470–1497. https://doi.org/10.1016/J.RSER.2016.03.036
45. Pons O, Aguado A, Fernández AI, Cabeza LF, Chimenos JM (2014) Review of the use of phase change materials (PCMs) in buildings with reinforced concrete structures. Mater Constr 64(315):e031–e031. https://doi.org/10.3989/MC.2014.05613
46. Alva G, Lin Y, Liu L, Fang G (2017) Synthesis, characterization and applications of microencapsulated phase change materials in thermal energy storage: a review. Energy Build 144:276–294. https://doi.org/10.1016/J.ENBUILD.2017.03.063
47. Fang Y (2022) A comprehensive study of phase change materials (PCMs) for building walls applications. Accessed 18 Dec 2022. [Online]. Available: https://kuscholarworks.ku.edu/handle/1808/5537
48. Vakhshouri AR, Vakhshouri AR (2019) Paraffin as phase change material. Paraffin Overview. https://doi.org/10.5772/INTECHOPEN.90487
49. Vérez D, Borri E, Zsembinszki G, Cabeza LF (2023) Thermal energy storage co-benefits in building applications transferred from a renewable energy perspective. J Energy Storage 58:106344. https://doi.org/10.1016/J.EST.2022.106344
50. Cui Y, Xie J, Liu J, Pan S (2015) Review of phase change materials integrated in building walls for energy saving. Procedia Eng 121:763–770. https://doi.org/10.1016/J.PROENG.2015.09.027

51. Al-Yasiri Q, Szabó M (2021) Selection of phase change material suitable for building heating applications based on qualitative decision matrix. Energy Convers Manag: X 12:100150. https://doi.org/10.1016/J.ECMX.2021.100150
52. Al-Absi ZA, Hafizal MIM, Ismail M, Awang H, Al-Shwaiter A (2022) Properties of PCM-based composites developed for the exterior finishes of building walls. Case Stud Constr Mater 16:e00960. https://doi.org/10.1016/J.CSCM.2022.E00960
53. Hawes DW, Banu D, Feldman D (1989) Latent heat storage in concrete. Solar Energy Mater 19(3–5):335–348. https://doi.org/10.1016/0165-1633(89)90014-2
54. Ismail ZZ, AL-Hashmi EA (2009) Recycling of waste glass as a partial replacement for fine aggregate in concrete. Waste Manag 29(2):655–659. https://doi.org/10.1016/J.WASMAN.2008.08.012
55. Ali Memon S, Yiu Lo T, Shi X, Barbhuiya S, Cui H (2013) Preparation, characterization and thermal properties of Lauryl alcohol/Kaolin as novel form-stable composite phase change material for thermal energy storage in buildings. Appl Therm Eng 59(1–2):336–347. https://doi.org/10.1016/J.APPLTHERMALENG.2013.05.015
56. Al-Yasiri Q, Szabó M (2021) Incorporation of phase change materials into building envelope for thermal comfort and energy saving: a comprehensive analysis. J Build Eng 36:102122. https://doi.org/10.1016/J.JOBE.2020.102122
57. Zhang X et al (2017) Shape-stabilized composite phase change materials with high thermal conductivity based on stearic acid and modified expanded vermiculite. Renew Energy 112:113–123. https://doi.org/10.1016/J.RENENE.2017.05.026
58. Ramakrishnan S, Wang X, Sanjayan J (2018) Thermal enhancement of paraffin/hydrophobic expanded perlite granular phase change composite using graphene nanoplatelets. Energy Build 169:206–215. https://doi.org/10.1016/j.enbuild.2018.03.053
59. Rathore PKS, Kumar Shukla S (2021) Improvement in thermal properties of PCM/Expanded vermiculite/expanded graphite shape stabilized composite PCM for building energy applications. Renew Energy 176:295–304. https://doi.org/10.1016/j.renene.2021.05.068
60. Al-Kaabi JJF (2024) Evaluating performance characteristics of recycled aggregate concrete: a study through experimentation. Herit Sustain Dev 6(2):557–570. https://doi.org/10.37868/HSD.V6I2.815
61. Kumar N, Rathore PKS, Sharma RK, Gupta NK (2023) Integration of lauric acid/zeolite/graphite as shape stabilized composite phase change material in gypsum for enhanced thermal energy storage in buildings. Appl Therm Eng 224:120088. https://doi.org/10.1016/J.APPLTHERMALENG.2023.120088
62. Yu H, Li C, Zhang K, Tang Y, Song Y, Wang M (2020) Preparation and thermophysical performance of diatomite-based composite PCM wallboard for thermal energy storage in buildings. J Build Eng 32. https://doi.org/10.1016/j.jobe.2020.101753
63. Zhu N, Wang S, Xu X, Ma Z (2010) A simplified dynamic model of building structures integrated with shaped-stabilized phase change materials. Int J Therm Sci 49(9):1722–1731. https://doi.org/10.1016/J.IJTHERMALSCI.2010.03.020
64. Zhang N, Yuan Y, Yuan Y, Li T, Cao X (2014) Lauric–palmitic–stearic acid/expanded perlite composite as form-stable phase change material: preparation and thermal properties. Energy Build 82:505–511. https://doi.org/10.1016/J.ENBUILD.2014.07.049
65. Zhou G, He J (2015) Thermal performance of a radiant floor heating system with different heat storage materials and heating pipes. Appl Energy 138:648–660. https://doi.org/10.1016/J.APENERGY.2014.10.058
66. Zhou XM (2009) Preparation and characterization of PEG/MDI/PVA copolymer as solid–solid phase change heat storage material. J Appl Polym Sci 113(3):2041–2045. https://doi.org/10.1002/APP.29923
67. Fallahi A, Guldentops G, Tao M, Granados-Focil S, Van Dessel S (2017) Review on solid-solid phase change materials for thermal energy storage: molecular structure and thermal properties. Appl Therm Eng 127:1427–1441. https://doi.org/10.1016/J.APPLTHERMALENG.2017.08.161

68. Sani AK, Olawoore IO, Singh RM (2021) Assessment of impregnating phase change materials into lightweight aggregates for development of thermal energy storage aggregate composites. Constr Build Mater 305:124683. https://doi.org/10.1016/J.CONBUILDMAT.2021.124683
69. Son YK, Ko DC, Kim BM (2015) Prediction of delamination and tearing during stamping of polymer-coated metal sheet. J Mater Process Technol 220:146–156. https://doi.org/10.1016/J.JMATPROTEC.2015.01.013
70. Cunha S, Leite P, Aguiar J (2020) Characterization of innovative mortars with direct incorporation of phase change materials. J Energy Storage 30:101439. https://doi.org/10.1016/J.EST.2020.101439
71. Suresh C, Kumar Hotta T, Saha SK (2022) Phase change material incorporation techniques in building envelopes for enhancing the building thermal comfort-a review. Energy Build 268:112225. https://doi.org/10.1016/J.ENBUILD.2022.112225
72. Lopez-Arias M, Francioso V, Velay-Lizancos M (2023) High thermal inertia mortars: new method to incorporate phase change materials (PCMs) while enhancing strength and thermal design models. Constr Build Mater 370:130621. https://doi.org/10.1016/J.CONBUILDMAT.2023.130621
73. Salunkhe PB, Shembekar PS (2012) A review on effect of phase change material encapsulation on the thermal performance of a system. Renew Sustain Energy Rev 16(8):5603–5616. https://doi.org/10.1016/J.RSER.2012.05.037
74. Chinnasamy V, Heo J, Jung S, Lee H, Cho H (2023) Shape stabilized phase change materials based on different support structures for thermal energy storage applications–a review. Energy 262:125463. https://doi.org/10.1016/J.ENERGY.2022.125463
75. Li Q et al (2022) Thermal performance and economy of PCM foamed cement walls for buildings in different climate zones. Energy Build 277:112470. https://doi.org/10.1016/J.ENBUILD.2022.112470
76. Bentz DP, Turpin R (2007) Potential applications of phase change materials in concrete technology. Cem Concr Compos 29(7):527–532. https://doi.org/10.1016/j.cemconcomp.2007.04.007
77. Farnam Y, Esmaeeli HS, Zavattieri PD, Haddock J, Weiss J (2017) Incorporating phase change materials in concrete pavement to melt snow and ice. Cem Concr Compos 84:134–145. https://doi.org/10.1016/J.CEMCONCOMP.2017.09.002
78. Ma B, Adhikari S, Chang Y, Ren J, Liu J, You Z (2013) Preparation of composite shape-stabilized phase change materials for highway pavements. Constr Build Mater 42:114–121. https://doi.org/10.1016/j.conbuildmat.2012.12.027
79. Sakulich AR, Bentz DP (2012) Increasing the service life of bridge decks by incorporating phase-change materials to reduce freeze-thaw cycles. J Mater Civ Eng 24(8):1034–1042. https://doi.org/10.1061/(ASCE)MT.1943-5533.0000381
80. Zhang P, Song L, Lu H, Wang J, Hu Y (2010) The influence of expanded graphite on thermal properties for paraffin/high density polyethylene/chlorinated paraffin/antimony trioxide as a flame retardant phase change material. Energy Convers Manag 51(12):2733–2737. https://doi.org/10.1016/j.enconman.2010.06.009
81. De Gracia A, Cabeza LF (2015) Phase change materials and thermal energy storage for buildings. Energy Build 103:414–419. https://doi.org/10.1016/j.enbuild.2015.06.007
82. Cabeza LF, Castellón C, Nogués M, Medrano M, Leppers R, Zubillaga O (2007) Use of microencapsulated PCM in concrete walls for energy savings. Energy Build 39(2):113–119. https://doi.org/10.1016/j.enbuild.2006.03.030
83. Silva T, Vicente R, Soares N, Ferreira V (2012) Experimental testing and numerical modelling of masonry wall solution with PCM incorporation: a passive construction solution. Energy Build 49:235–245. https://doi.org/10.1016/j.enbuild.2012.02.010
84. Eddhahak-Ouni A, Drissi S, Colin J, Neji J, Care S (2014) Experimental and multi-scale analysis of the thermal properties of Portland cement concretes embedded with microencapsulated Phase Change Materials (PCMs). Appl Therm Eng 64(1–2):32–39. https://doi.org/10.1016/j.applthermaleng.2013.11.050

85. Kheradmand M, de Aguiar JB, Azenha M (2014) Estimation of the specific enthalpy–temperature functions for plastering mortars containing hybrid mixes of phase change materials. Int J Energy Environ Eng 5(1). https://doi.org/10.1007/S40095-014-0081-9
86. Özrahat E, Ünalan S (2017) Thermal performance of a concrete column as a sensible thermal energy storage medium and a heater. Renew Energy 111:561–579. https://doi.org/10.1016/j.renene.2017.04.046
87. Lecompte T, Le Bideau P, Glouannec P, Nortershauser D, Le Masson S (2015) Mechanical and thermo-physical behaviour of concretes and mortars containing phase change material. Energy Build 94:52–60. https://doi.org/10.1016/j.enbuild.2015.02.044
88. Gowreesunker BL, Tassou SA (2013) Effectiveness of CFD simulation for the performance prediction of phase change building boards in the thermal environment control of indoor spaces. Build Environ 59:612–625. https://doi.org/10.1016/j.buildenv.2012.10.004
89. Duraković B (2020) PCM-based glazing systems and components. In: Green energy and technology. Springer, pp 89–119. https://doi.org/10.1007/978-3-030-38335-0_5
90. Duraković B, Mešetović S (2019) Thermal performances of glazed energy storage systems with various storage materials: an experimental study. Sustain Cities Soc 45:422–430. https://doi.org/10.1016/j.scs.2018.12.003
91. Durakovic B, Torlak M (2017) Experimental and numerical study of a PCM window model as a thermal energy storage unit. Int J Low-Carbon Technol 12(3):272–280. https://doi.org/10.1093/ijlct/ctw024
92. Findik F, Ermiş K (2020) Thermal energy storage. Sustain Eng Innov 2(2):66–88. https://doi.org/10.37868/sei.v2i2.115
93. Amran YHM, Farzadnia N, Ali AAA (2015) Properties and applications of foamed concrete; a review. Constr Build Mater 101:990–1005. https://doi.org/10.1016/J.CONBUILDMAT.2015.10.112
94. Ramamurthy K, Kunhanandan Nambiar EK, Indu Siva Ranjani G (2009) A classification of studies on properties of foam concrete. Cem Concr Compos 31(6):388–396. https://doi.org/10.1016/J.CEMCONCOMP.2009.04.006
95. Huo T et al (2018) China's energy consumption in the building sector: a statistical yearbook-energy balance sheet based splitting method. https://doi.org/10.1016/j.jclepro.2018.02.283
96. Sci-Hub|Integrating phase change materials in construction materials: critical review. Constr Build Mater 217:36–49. https://doi.org/10.1016/j.conbuildmat.2019.05.064. Accessed 23 Nov 2022
97. Alqallaf HJ, Alawadhi EM (2013) Concrete roof with cylindrical holes containing PCM to reduce the heat gain. Energy Build 61:73–80. https://doi.org/10.1016/J.ENBUILD.2013.01.041
98. Anupam BR, Sahoo UC, Rath P (2020) Phase change materials for pavement applications: a review. Constr Build Mater 247:118553. https://doi.org/10.1016/J.CONBUILDMAT.2020.118553
99. Yang M, Zhang X, Zhou X, Liu B, Wang X, Lin X (2021) Research and exploration of phase change materials on solar pavement and asphalt pavement: a review. J Energy Storage 35:102246. https://doi.org/10.1016/J.EST.2021.102246
100. Methode Kalombe R, Sobhansarbandi S, Kevern J (2023) Low-cost phase change materials based concrete for reducing deicing needs. Constr Build Mater 363:129129. https://doi.org/10.1016/J.CONBUILDMAT.2022.129129
101. Shaikh PH, Nor NBM, Sahito AA, Nallagownden P, Elamvazuthi I, Shaikh MS (2017) Building energy for sustainable development in Malaysia: a review. Renew Sustain Energy Rev 75:1392–1403. https://doi.org/10.1016/J.RSER.2016.11.128
102. Shukla N, Fallahi A, Kosny J (2012) Performance characterization of PCM impregnated gypsum board for building applications. Energy Procedia 30:370–379. https://doi.org/10.1016/J.EGYPRO.2012.11.044
103. Yang Y, Shen Z, Wu W, Zhang H, Ren Y, Yang Q (2022) Preparation of a novel diatomite-based PCM gypsum board for temperature-humidity control of buildings. Build Environ 226:109732. https://doi.org/10.1016/J.BUILDENV.2022.109732

104. Li Y et al (2022) Processing wood into a phase change material with high solar-thermal conversion efficiency by introducing stable polyethylene glycol-based energy storage polymer. Energy 254:124206. https://doi.org/10.1016/J.ENERGY.2022.124206
105. Farid MM, Husian RM (1990) An electrical storage heater using the phase-change method of heat storage. Energy Convers Manag 30(3):219–230. https://doi.org/10.1016/0196-8904(90)90003-H
106. Wahid MA, Hosseini SE, Hussen HM, Akeiber HJ, Saud SN, Mohammad AT (2017) An overview of phase change materials for construction architecture thermal management in hot and dry climate region. Appl Therm Eng 112:1240–1259. https://doi.org/10.1016/J.APPLTHERMALENG.2016.07.032
107. Pirdavari P, Hossainpour S (2020) Numerical study of a Phase Change Material (PCM) embedded solar thermal energy operated cool store: a feasibility study. Int J Refrig 117:114–123. https://doi.org/10.1016/J.IJREFRIG.2020.04.028
108. Wang X, Yu H, Li L, Zhao M (2016) Experimental assessment on a kind of composite wall incorporated with shape-stabilized phase change materials (SSPCMs). Energy Build 128:567–574. https://doi.org/10.1016/J.ENBUILD.2016.07.031
109. Jia C, Geng X, Liu F, Gao Y (2021) Thermal behavior improvement of hollow sintered bricks integrated with both thermal insulation material (TIM) and Phase-Change Material (PCM). Case Stud Therm Eng 25:100938. https://doi.org/10.1016/J.CSITE.2021.100938
110. Principi P, Fioretti R (2012) Thermal analysis of the application of PCM and low emissivity coating in hollow bricks. Energy Build 51:131–142. https://doi.org/10.1016/J.ENBUILD.2012.04.022
111. Xu B, Li Z (2013) Paraffin/diatomite composite phase change material incorporated cement-based composite for thermal energy storage. Appl Energy 105:229–237. https://doi.org/10.1016/J.APENERGY.2013.01.005
112. Dallaire J, Adeel Hassan HM, Bjernemose JH, Rudolph Hansen MP, Lund I, Veje CT (2022) Performance analysis of a dual-stack Air-PCM heat exchanger with novel air flow configuration for cooling applications in buildings. Build Environ 223:109450. https://doi.org/10.1016/J.BUILDENV.2022.109450
113. Royon L, Karim L, Bontemps A (2013) Thermal energy storage and release of a new component with PCM for integration in floors for thermal management of buildings. Energy Build 63:29–35. https://doi.org/10.1016/J.ENBUILD.2013.03.042
114. Du Y et al (2019) Laboratory investigation of phase change effect of polyethylene glycolon on asphalt binder and mixture performance. Constr Build Mater 212:1–9. https://doi.org/10.1016/j.conbuildmat.2019.03.308
115. Findik F (2025) Sustainable composite materials: applications and future. Sustain Eng Innov 7(2):415–424. https://doi.org/10.37868/SEI.V7I2.ID504
116. Al-Absi ZA, Isa MHM, Ismail M (2020) Phase change materials (PCMs) and their optimum position in building walls. Sustainability (Switzerland) 12(4). https://doi.org/10.3390/SU12041294
117. Hemmati R, Saboori H (2016) Short-term bulk energy storage system scheduling for load leveling in unit commitment: modeling, optimization, and sensitivity analysis. J Adv Res 7(3):360–372. https://doi.org/10.1016/J.JARE.2016.02.002
118. Sci-Hub|The effect of ultraviolet coating on containment and fire hazards of phase change materials impregnated wood structure. J Energy Storage 32:101727. https://doi.org/10.1016/j.est.2020.101727. Accessed 09 Dec 2022
119. Gencel O et al (2022) Properties of eco-friendly foam concrete containing PCM impregnated rice husk ash for thermal management of buildings. J Build Eng 58:104961. https://doi.org/10.1016/J.JOBE.2022.104961
120. Tarasov AS, Kearsley EP, Kolomatskiy AS, Mostert HF (2015) Heat evolution due to cement hydration in foamed concrete. Mag Concr Res 62(12):895–906. https://doi.org/10.1680/MACR.2010.62.12.895
121. Umar H et al (2018) Mechanical properties of concrete containing beeswax/dammar gum as phase change material for thermal energy storage. AIMS Energy 3(3):521–529. https://doi.org/10.3934/ENERGY.2018.3.521

122. Al-Absi ZA, Hafizal MIM, Ismail M (2022) Experimental study on the thermal performance of PCM-based panels developed for exterior finishes of building walls. J Build Eng 52:104379. https://doi.org/10.1016/J.JOBE.2022.104379
123. Sun X, Medina MA, Lee KO, Jin X (2018) Laboratory assessment of residential building walls containing pipe-encapsulated phase change materials for thermal management. Energy 163:383–391. https://doi.org/10.1016/J.ENERGY.2018.08.159
124. Al-Yasiri Q, Szabó M (2021) Effect of encapsulation area on the thermal performance of PCM incorporated concrete bricks: a case study under Iraq summer conditions. Case Stud Constr Mater 15:e00686. https://doi.org/10.1016/J.CSCM.2021.E00686
125. Shaik S, Arumugam C, Shaik SV, Arıcı M, Afzal A, Ma Z (2022) Strategic design of PCM integrated burnt clay bricks: potential for cost-cutting measures for air conditioning and carbon dioxide extenuation. J Clean Prod 375:134077. https://doi.org/10.1016/J.JCLEPRO.2022.134077
126. Kong X, Jiang L, Yuan Y, Qiao X (2022) Experimental study on the performance of an active novel vertical partition thermal storage wallboard based on composite phase change material with porous silica and microencapsulation. Energy 239:122451. https://doi.org/10.1016/J.ENERGY.2021.122451
127. Bake M, Shukla A, Liu S (2021) Development of gypsum plasterboard embodied with microencapsulated phase change material for energy efficient buildings. Mater Sci Energy Technol 4:166–176. https://doi.org/10.1016/J.MSET.2021.05.001
128. Jeong SG, Wi S, Chang SJ, Lee J, Kim S (2019) An experimental study on applying organic PCMs to gypsum-cement board for improving thermal performance of buildings in different climates. Energy Build 190:183–194. https://doi.org/10.1016/J.ENBUILD.2019.02.037
129. Molinari C et al (2021) Effect of scale-up on the properties of PCM-impregnated tiles containing glass scraps. Case Stud Constr Mater 14:e00526. https://doi.org/10.1016/J.CSCM.2021.E00526
130. Hatamleh RI, Abu-Hamdeh NH, Khoshaim A, Alzahrani MA (2022) Using phase change material (PCM) to improve the solar energy capacity of glass in solar collectors by enhancing their thermal performance via developed MD approach. Eng Anal Bound Elem 143:163–169. https://doi.org/10.1016/J.ENGANABOUND.2022.06.010
131. Mohammed AM, Elnokaly A, Aly AMM (2021) Empirical investigation to explore potential gains from the amalgamation of phase changing materials (PCMs) and wood shavings. Energy Built Environ 2(3):315–326. https://doi.org/10.1016/J.ENBENV.2020.07.001
132. Chen H, Xuan J, Deng Q, Gao Y (2022) WOOD/PCM composite with enhanced energy storage density and anisotropic thermal conductivity. Prog Nat Sci Mater Int 32(2):190–195. https://doi.org/10.1016/J.PNSC.2022.01.002
133. Nebrida JA, Gomba FE (2023) Sustainable construction strategies for building construction projects in the Kingdom of Bahrain: a model. Sustain Eng Innov 5(1):31–47. https://doi.org/10.37868/SEI.V5I1.ID193

Chapter 8
Renewable Energy Systems and Application

Abstract Buildings are considered as a key lever in reducing the reliance on non-renewable energy sources and greenhouse gasses emission. Renewable technologies are enablers for net-zero energy buildings and reduce their reliance on electrical grids. This chapter examines a range of these technologies including those already widely available as well as those in their early development. The results indicate that, although continued technological advancement is desirable, the current state of renewable technologies is significant to substantially reduce greenhouse gases emissions and increase societal resilience.

Keywords Zero-emission technologies · Carbon emissions · Renewable energy source · Efficiency · Energy storage · Carbon capture · Carbon removal

8.1 Introduction

Anthropogenic climate change has been one of the largest societal concerns of the twenty-first century, with research consistently pointing towards a doomsday scenario for the planet if nothing is done to resolve this issue. Per IEA, the total amount of CO_2 emissions in 2021 reached 36.6 Gt, with the current emissions trajectory leading to a potential rise of 2.5 °C in global average temperatures by 2100 [1]. Although there has been much positive movement towards renewable energy sources, the global effort needed to avoid negative environmental consequences must be much higher than what is currently being done. The unfortunate geo-political events that shaped the past few years have halted some of the progress that was being made to end our reliance on fossil fuels and taken away attention and resources from the issue of climate change. Buildings have naturally been considered as an important part of this effort to lower global carbon footprint. Although it is crucial to look at the broader scale of how energy is used and produced, many of the solutions will be found on the micro level of technologies that can be applied individually to buildings which will help lower building energy demand. These technologies will enable buildings to become more sustainable to the point where they could be considered

B. Duraković, *Zero Energy Buildings*, Green Energy and Technology,
https://doi.org/10.1007/978-3-032-16027-0_8

as net-zero energy buildings (NZEB). The technologies that enable net-zero energy buildings are the topic of this chapter.

Although "net-zero" is quite a broad term that can be used and applied in many different fields, the concept of Net-Zero Technology as it pertains to sustainability is to provide a balance between the amount energy that is needed for building operations and the amount of energy that the building generate. It can also refer to the balance between greenhouse gas produced and the amount removed from the atmosphere [2]. Using Net-Zero Technology allows buildings to be self-sufficient instead of relying on the local energy grid. This results in more reliant buildings and in less greenhouse gas emissions since. Producing non-renewable energy almost always requires a very large and expensive setup, so naturally these sources will be limited to energy systems that provide energy to a large area and/or population. In the efforts to reduce the use of fossil fuels, zero-emission technologies will replace our existing non-renewable energy system. Many zero-emission technologies are already in use while the others are still only theoretical or in the testing phase. Technologies such as advanced solar energy, improved biomass utilization for recycling combined with carbon capture and storage (CCS), and advanced nuclear technologies [3] are anticipated to eliminate reliance on non-renewable energy sources in the decades ahead [4].

According to ASHRAE,[1] zero-energy building (ZEB) is a building that on annual basis uses no more energy than is provided by the building's on-site renewable energy resources. NZEB/ZEB can become independent from the fluctuation of energy prices and availability by investing in their own renewable energy systems which have a predictable expected outcome, meaning that these buildings are fully sustainable. According to research done in 2011 by Politecnico di Milano, a ZEB becomes more economical in comparison to a similar building within a short period of time (15–20 years) [5]. It can be assumed that with the technological advancement of renewable energy technologies over the last decade, the economic advantage of a net-zero energy building has become higher.

8.2 Renewable Energy Technologies

There is a growing range of technologies available to companies and individuals interested in achieving zero emissions in their buildings. Some of these technologies have been available since the latter half of the twentieth century, most of them have made significant improvements and advancements over the decades. Other technologies are still in the early stages of development or under theoretical concepts or ideas being tested. While many of the new concepts that promise to revolutionize energy, infrastructure is under focus of researchers, it is important to note that even with the current state of technology, zero-emission buildings are often achievable. Renewable energy sources have potential to meet current energy demand multiple times [6], making concerns about energy scarcity largely irrelevant.

[1] American Society of Heating, Refrigerating and Air-Conditioning Engineers.

8.2.1 Solar Energy Technologies

Solar energy holds the greatest potential for replacing a significant portion of the current energy infrastructure, as it is an abundant resource that has been effectively harnessed for many decades. Therefore, solar energy can be converted either into heat or directly into electricity. The oldest and the most common technology for using solar energy is solar collectors that collect and deliver solar energy to via heat transfer fluid into buildings. Common types of solar collectors commercially available are flat plate collectors and evacuated tube [7].

Flat plate collectors have low cost, and they are relatively easy to manufacture and install that resulted in wide application [8]. They are commonly applied in industrial and domestic buildings. Evacuated tube collectors are constructed of one or several rows of tubes connected to the manifold. Each tube consists of an outer thick glass tube and a thinner inner glass tube coated with a solar energy-absorbing material. The vacuum inside the tubes minimizes heat loss, making evacuated tube collectors more efficient than flat plate collectors. Evacuated solar tubes are primarily classified into two main types: single-walled glass and Dewar tubes, with several subcategories within these two groups [8].

Some of the main parameters for solar collectors are roof or ground mounting, high or low temperature range, flat plate or evacuated tube collector through which heat transfer fluid circulates [9]. These collectors require non-sustainable methods and materials for construction, leading to an additional carbon footprint that ought to be offset during its lifetime, so the usage of more sustainable materials and construction/installation methods is highly encouraged. In the case of solar thermal collectors, it has been found that sustainable materials such as stubble fiber and rice husk can be used for insulation as a replacement for inorganic materials [10]. Another type of cogeneration system exists which uses solar energy, a solar photovoltaic thermal (PV/T) system, which produces both electricity and heat [11].

Conversion of solar energy into electricity can be done using photovoltaic (PV), thermophotovoltaic (TPV), concentrated solar power (CSP), and concentrated photovoltaic (CPV) technologies. PV systems directly convert sunlight into electricity using semiconductor materials. TPV panels generate electricity by utilizing radiant heat differentials through the emission of photons. CSP employs mirrors or lenses to focus sunlight onto a fluid, which is then heated to drive a turbine and generate electricity. Each of these technologies offers unique approaches to harnessing solar energy, contributing to a diverse range of renewable energy solutions. CPV systems utilize optical elements like Fresnel lenses or parabolic mirrors to focus sunlight onto high-efficiency multi-junction solar cells, significantly enhancing energy conversion rates [12]. These systems require precise dual axis tracking to maintain optimal alignment with the sun, ensuring maximum light concentration. CPV technology is particularly effective in regions with high direct sunlight, offering a compact and scalable solution for solar energy generation [13].

The PV systems can be classified into three generations. The first generation is Single crystalline and multi crystalline silicate material solar cells. The second generation comprises mainly thin film PV cells such as amorphous silicon, cadmium sulfide (CDS) and cadmium telluride (CDTe). The third generation includes other silicon-free devices such as Perovskite PV cells, dye-sensitize PV cells, and quantum dot PV cells [14]. PV cells need to fulfil some essential conditions to be widespread including low cost, high efficiency and long service life. The PV cell made through crystalline materials can meet all these requirements of energy; electricity and thermal [15].

The current efficiency of PV panels ranges from 20% up to approximately 40%, for TPV and CSP panels, as well as for concentrated photovoltaic panels (CPV), although it is possible for this efficiency to be increased up to approximately 60% [16]. The most used type of technology is PV panels. One of the big benefits of this technology is that it is relatively cheap compared to its alternatives, not as geographically restricted as some other renewable technologies, and can be installed on a very small scale such as a roof of a house, as well as much larger scale developments such as "solar farms". These advantages have resulted in PV panels being a very popular choice for renewable energy generation with fastest growth [17]. However, a frequently cited PV panels drawback in the literature is their relatively short lifetime of 25–30 years [18], after which there is no clear use for these panels, leading to lots of potential waste if this issue is not resolved. It is because of the long history and success of solar panels that we can learn a lot from how even renewable energy sources can have significant negative side effects.

8.2.2 Wind Energy Technologies

Wind Energy has become an essential part in fighting climate change, and large wind turbines have been installed across the globe, strategically placed where they can harvest the highest amount of energy. Similarly to solar energy, wind energy can be harvested on both a large scale and a small scale. While wind turbines can reach heights of well over 100 m, there are also many commercially available small turbines that can be placed on a building's roof or in a garden. The International Electrotechnical Commission defines small wind turbines (SWT) as wind turbines with maximum 200 m^2 rotor swept area and generating electricity at a voltage below 1000 V a.c. or 1500 V d.c. for both on-grid and off-grid applications [19]. Wind profiles, installation height, land use, characteristics of the turbine are some of the main parameters that determine the efficiency of a wind turbine [20]. The most important differentiation of small wind turbines is the Horizontal Axis Wind Turbine (HAWT) and Vertical Axis Wind Turbine (VAWT). A good site, low impact materials and a reliable construction allow SWTs to reduce the environmental impact of electricity production [21]. Efficiency ranges for different types of wind turbines are shown in Table 8.1.

Wind Energy (WE) system has preceded other types of renewable energy systems. Wind turbines are among the most important items that have taken years to develop.

Table 8.1 Wind turbines efficiency ranges

Turbine type	Efficiency (%)	Details	Source
Small horizontal axis wind turbine	30–40	Efficiency varies based on location, wind profile, and installation height; higher efficiency in areas with consistent wind speeds	[22–25]
Small vertical axis wind turbine	15–25	Lower efficiency due to design constraints, but works well in urban settings with variable wind directions	[22–25]
Large horizontal axis wind turbine	40–50	Higher efficiency due to larger rotor diameter and higher tip speeds; commonly used in wind farms and commercial applications	[23–26]

This is evident through expanding size or enhancing the capacity of existing structures. The other signal is economically rising power generated. A huge increase in the number of installed WE systems imposed higher demands for connection to improve the stability of such systems [27]. Relative increase in the level of complexity calls for better energy control mechanisms to steer the power produced by wind turbines and make it available for consumers sufficiently and reliably. Figure 8.1 shows detailed trends in turbine capacity and hub height over the last three decades.

Figure 8.1 demonstrates in detail how wind turbine performance has improved, and technology has advanced over time. While the increase in hub height provides access to stronger and more steady wind flows, the increase in turbine capacity suggests a larger potential for energy production. Design optimizations and efficiency improvements are reflected in changes in specific power. These advancements make

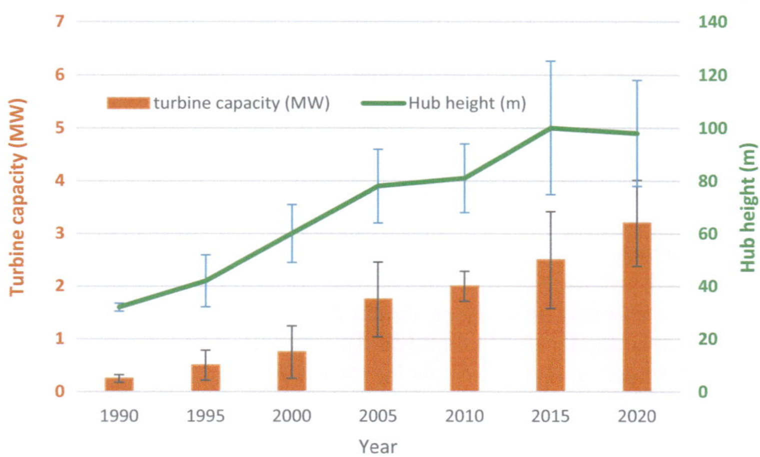

Fig. 8.1 Evolution of Europe's onshore wind turbines over the last three decades (based on data presented in [28])

it possible for wind energy systems to contribute to energy grids in a more dependable and sustainable manner.

Most wind turbines had a capacity of 2 MW up to 2005 till 2013, and from 2014 and up to now, it has stepped up to 3/3.5 MW. Also, the tower height grew more than two times since 1990 to 100 m and more. The blade length and its operating range cut-in as well as cut out both have been increased [28]. Thus, the nacelle and tower make a very large contribution to a small-scale wind turbine [29]. The recent technological enhancement in the field of wind turbine assembly has stretched up limits of generation. The size of the unit has grown to reach 3.00 MW and above, which has a significant position in electrical grid. The large wind market has received much more strength from governmental support than the small-scale market which is still limited. Hence, the expansion of small-scale turbines requires that the cost of installation should come down and that more encouragement should be given by the government [30].

8.2.3 Biomass and Biogas Technologies

Biomass and biogas can be converted into clean energy that produces sustainable greenhouse gas emissions and no water or ground pollution. Different forms of bio fueled generated through biomasses include bioethanol, biodiesel, biogas, bio butanol, bio-oil and bio hydrogen [31]. Biomass is available in three basic forms of the substance: And divided into solid, liquid and gaseous, which can concurrently be subclassified as primary materials and by-products [32, 33]. As part of the basic functioning of society, we produce biomass waste which include food waste (FW), wood waste (WW), agricultural waste (AW), sewage sludge waste (SSW) and municipal solid waste (MSW) [34]. This waste usually ends up as composting, fertilizer, foddering, and landfilling, which produces a negative effect on the environment. Biomass waste can be used instead to produce clean energy. The types of biomasses are quite varied and involve multiple methods of converting biomass into energy. The size of biomass digesters is contingent upon various factors, including the location, the number of households they serve, and the quantity of available substrate daily [35]. The most common way that biomass is used for energy is in developing countries where firewood is used for cooking and heating. This usage is not without its drawbacks, as it results in local air pollution. While the release of CO_2 is an inevitable part of many of the common methods of using biomass, new technologies are being developed that would capture this CO_2 and make this energy production processes considered as environmentally friendly. Table 8.2 provides a concise breakdown of the different types of biomass and biogas technologies, their efficiencies, descriptions, and the most appropriate contexts for their use.

Biogas and biomass technologies are essential for sustainable energy production by converting organic waste into renewable energy. These technologies provide clean energy and support rural communities and have substantial efficiencies. Using these

Table 8.2 Overview of biomass and biogas technologies

Technology	Efficiency	Description	Applicatoin	Source
Biogas from organic waste	60–70	Anaerobic digestion of organic materials produces biogas (methane) and digestate	Rural energy needs, waste treatment, and small-scale power generation	[36–39]
Biogas from agricultural waste	60–70	Conversion of agricultural residues (e.g., manure, crop waste) into biogas	Agricultural communities for cooking, heating, and electricity	[36–38, 40]
Biomass combustion	70–85	Direct combustion of organic materials (wood, crop residues) to produce heat	District heating, industrial processes, and residential heating	[36, 41]
Biomass gasification	75–85	Thermal conversion of biomass into syngas for electricity and heat production	Combined heat and power (CHP) systems, industrial applications	[36, 41, 42]
Biomass pyrolysis	60–70	Decomposition of biomass in the absence of oxygen to produce bio-oil, syngas, and charcoal	Biofuel production, chemical feedstocks, and soil enhancement	[36, 41, 42]

technologies sustainability is promoted by reducing waste, lowering carbon footprints, and decreasing reliance on fossil fuels. Challenges like feedstock consistency and high initial costs must be addressed.

Biomass fuels account for 10–14% of the total consumed energy; 40% of the energy produced in urban settings and 90% in rural settings. Biomass is used for power generation through the conversion of the energy stored in organic materials into electricity and for heat production by using the technological process of conversion of biomass to heat energy and biofuels. These processes remain dynamic and thus biomass is a highly sustainable and viable means of generating renewable energy for the future [43].

8.3 Geothermal Energy

Geothermal Energy (GTE) is one of the most promising and fastest growing renewable energy sources. Due to its renewable nature, it is sustainable, and its supply is stable compared to other non-renewable resources. Being a dependable energy solution, it serves considerable interconnectedness to minimizing the dependency on fossil fuel and promoting the utilization of renewable energy. GTE is a significant role in countries like Iceland and El Salvador or where it generates a good portion of their electricity supply.

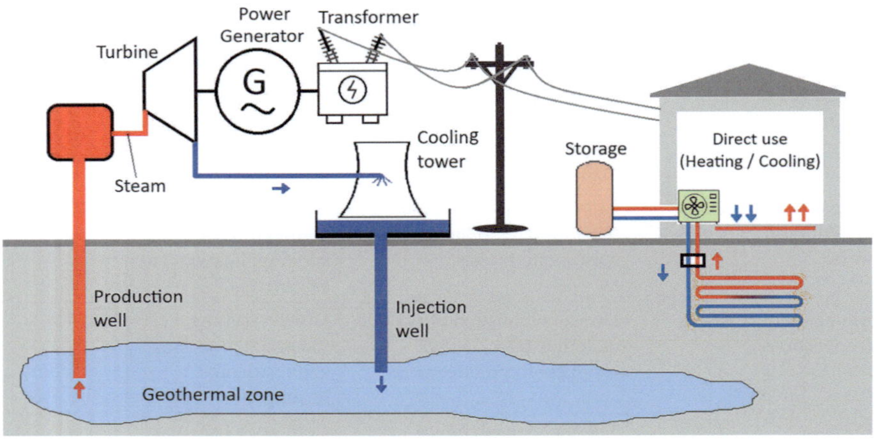

Fig. 8.2 Dual-use geothermal system: electrification and thermal applications in future energy mix

GTE is a standalone constantly available energy resource throughout the year irrespective of other energy sources. It offers a big share for power in Iceland where it supplies 27% of power and in El Salvador; 26%. It is a new RER which originates from the heat of the earth produced by molten magma and by radioactive decay of elements. Based on the system classification method it can be divided into shallow GTE, underground thermal water and hot dry rock resources. Figure 8.2 represents an integrated geothermal energy system, depicting subsurface reservoirs, production and reinjection wells, and the dual pathways to both electricity generation and direct heating applications.

Shallow surface GTE is low temperature and largely confined to the soil, gravel and water within the depth of up to 200 m from the surface. Thermal water with a temperature of more than 25 °C in a depth of less than 4000 m. Hot dry rock resources are at a temperature greater than 150 °C situated in depths between 3 and 10 km [44]. According to the details obtained from the survey, GTE mainly applied for generating electricity and heat. Consequently, geothermal energy provides nations with a unique opportunity to advance their green power generation initiatives through its dual capacity for direct electrification and heating. As technological advancements continue, the global share of geothermal energy in total energy consumption is projected to rise, establishing it as a cornerstone in the future renewable energy landscape [45].

8.3.1 Efficient HVAC Systems Technologies

Heating, ventilation, and air conditioning (HVAC) systems are an unavoidable part of a modern buildings and they are one of the largest consumers of energy. Trying to reduce the level of energy consumption in the world will require a close examination of the current state of HVAC systems and figuring out more efficient ways to cool and heat buildings. Among the most promising technologies for energy efficiency improvement have been heat pumps [46]. An efficient ventilation system is crucial in retaining thermal comfort as well as avoiding unnecessary energy losses. Most often ventilation systems will lose efficiency because of faults that are present either because of poor installation or lack of maintenance and upkeep [47]. The energy efficiency and amount of usage of a HVAC system will largely depend on the geographical location, the surroundings of the building, as well as its shape and materialization. Additionally, the way in which the system is operated can significantly increase or decrease its efficiency. The operation of HVAC systems tends to be suboptimal, potentially leading to an energy loss of 15–30% when the usage of HVAC is applied poorly [48].

Heat pump technologies serve as a replacement for traditional cooling and heating systems by using the air, water or Earth's temperature to reduce/increase the interior temperature. These systems have existed in different forms and have been commercially available for over five decades [49]. Unlike solar panels, this technology is a lot more expensive to implement, therefore careful consideration of all factors should be contemplated before making the decision to install such a system. Table 8.3 provide an overview of various types of heat pumps and their efficiencies based on literature.

Referring to Table 8.3, it is observed that different heat pump types and their typical efficiencies vary. Ground Source Heat Pumps generally exhibit the highest Coefficient of Performance (COP), ranging from 4.0 to 6.0, due to the stable underground temperatures they utilize, making them ideal for colder climates. GHPs come in two basic configurations: ground-coupled (closed loop) systems, which are installed in the ground, and groundwater (open loop) systems, which are installed in wells and lakes [7].

Table 8.3 Heat pumps and their coefficient of performance

Type of heat pump	Efficiency (COP)	Description	Best use	Source
Air source heat pump	3.0–5.0	Extracts heat from outdoor air; efficiency decreases in cold weather	Residential heating and cooling, moderate climates	[50, 51]
Ground source heat pump	4.0–6.0	Extracts heat from the ground; more efficient in cold climates	Residential and commercial heating in colder climates	[52, 53]
Water source heat pump	4.0–5.5	Uses heat from a nearby water source	Commercial and residential applications near water bodies	[53]
Air-to-water heat pump	3.0–4.5	Transfers heat to water for heating and hot water production	Residential heating, especially with underfloor heating systems	[50]

Air Source Heat Pumps and Air-to-Water Heat Pumps offer slightly lower but still significant COP ranging from 3.0 to 5.0, and they are commonly used in residential heating and cooling. Water Source Heat Pumps with COP between 4.0 and 5.5, are particularly suitable for buildings located near water sources. This variety offers the selection of the most appropriate and energy-efficient system based on climatic conditions, energy sources, and building needs. The climate, the site of the building, cost of traditional heating and cooling, potential tax benefits, labor and equipment costs among others are parameters that ought to be considered.

8.3.2 *Energy Storage Technologies*

As the supply and the demand for energy are constantly fluctuating, energy storage is a mandatory part of an energy production and supply system. Although energy can be stored in large-scale facilities, it has proven to be a useful tool on a smaller scale as well. Many of the technologies that have been mentioned above are more efficient with the application of an energy storage system (ESS). The importance of such a system only increases as the reliance on renewable energy increases since the availability of renewable energy tends to be considerably more volatile [54]. Multiple energy storage systems exist, with the main variables being size, cost and lifetime. Latent heat energy storage system (LHESS) [55], superconducting magnetic ESS (SMESS), supercapacitor (SC), lithium-ion and other battery types, compressed air energy storage system (CAESS), and flywheel energy storage system (FESS) are the most common ways to store energy [56]. Additionally, a photovoltaic, pumped thermal energy storage system (PV-PTES) applied on a small scale has been shown to decrease the dependence on the public grid by 2–28% and increase self-sustainability by 1–14% [57]. Energy storage capacity will have an average annual growth rate of 27% throughout this decade [58]. Energy Storage has become a key component of lowering our dependence on fossil fuels, by allowing us to reduce the volatility of renewable energy. ESS are discussed in detail in Chap. 5 of this book.

8.3.3 *Combined Heat and Power Technology*

Also known as cogeneration systems, CHP systems produce electricity while simultaneously providing useful thermal energy for heating purposes. Since traditional power generation is inherently wasteful, usually only generating approximately 30% useful energy from their primary source. Compared to central power stations and boilers, CHP can use up to 35% less primary energy [59]. Biomass and biogas can be used to supply the CHP system, allowing the building to avoid using fossil fuel for heating. These systems are most often used in buildings with large heating requirements or long heating seasons and are at this point more successful on larger scales, or as central plants to supply a group of buildings, but the market for small scale

CHP systems has been growing significantly [60]. Research has shown that a hybrid system involving Photovoltaics and CHP can considerably expand the potential of solar energy while reducing energy waste and lessen some of the faults of PV [61].

CHP systems are applicable for building/district heating, cooling and electricity generation. Waste heat from the turbine (or district plant) is used to run an absorption chiller. Building cooling/heating is achieved through the distribution of heat transfer from absorption chillers to users [62]. The CHP system is made up of a gas generator set, a wind turbine, a solar heater, an absorption heat pump unit, a ground source heat pump unit, a refrigeration unit, and other auxiliary equipment [63]. CCHP systems may all be combined in a variety of ways, and these combinations are routinely studied to meet thermodynamic and economic requirements. The sliding pressure operating system with the constant temperature gas storage tank has the maximum exergy efficiency and yearly profit [64]. As a result of their well-studied performance, CCHP systems are becoming increasingly popular. Natural gas use can be slightly decreased, resulting in savings in energy and a reduction in emissions. The CCHP system can be researched from three perspectives including energy, economics, and environmental performance, due to its relatively high cost [65].

8.4 Sustainable Technologies Implementation

The application of these technologies involves careful consideration of the factors that can either apply broadly to technology or specifically to building. Broadly speaking, every stated technology will have its limitations and framework within which it works best. Often multiple technologies will be grouped to get better results, or a piece of technology will require additional systems to properly function. The most common instance of this sort of technology pairing is using technologies which experience significant fluctuations in their energy output with batteries which can store energy and allow the user to have consistent energy availability [66]. The building itself will have unique parameters which can make it a good candidate for one technology, while disqualifying it from another one. The geographical location, regulations and building code, cost of materials and construction, the availability of net-zero technologies, and many other parameters will determine the success of a ZEB project. Other times, the technology might not be appropriately designed for certain regions, making it unusable, such as a different inverter which prohibits a system from being connected to the grid [67]. Although the application of these technologies is essential, the design of the buildings where they are applied is equally important. A high-quality building envelope can go a long way in reducing energy waste, significantly improving the energy performance of the building [68]. In fact, the need for these technologies can be limited or even eliminated with a good enough envelope, with the best example being Passive Houses, which are certified for reaching all criteria to achieve the Passive House Standard [69]. Unfortunately, such a standard is prohibitively expensive and complex for most construction, but there are still ways one can improve the building envelope without too high of an additional cost or complication of the construction process. The costs do not only extend to the envelope, and as discussed in some of the previous paragraphs, net-zero technologies all have varying costs. The value proposition of some of these technologies is too low at this moment, making them unattractive to many consumers, while others have seen a decrease in costs over time as their implementation grew so they have become quite feasible for most budgets. With time and further development, most of these systems will see a significant decrease in price, making them more approachable and viable. Figure 8.3 show photovoltaic system prices reduction over time. The diagram is generated based on data from provided in [70].

The main idea of ZEBs is to ensure long-term sustainability, hence it is important to calculate not the cost of installation, the immediate effects, etc. but additionally to look at the future. What are the costs going to be in the coming decades? How often will the systems need to be serviced or replaced? What external factors can impact the systems over time? These questions are answered by what is known as a Life Cycle Assessment (LCA), in which the whole life cycle of a system, from its production to its disposal, is taken into consideration. An LCA can significantly simplify decision making, since it gives a much clearer picture of the whole process of a technology's implementation [66]. Below in Table 8.4 are summarized all technologies discussed above with the results of their implementation.

Implementing these technologies successfully relies on a complicated set of considerations, some of which can be dealt with locally, while others require the involvement of the broader populace, scientific community, and often policy makers. All these factors and many others will influence the feasibility of NZEBs. As more buildings implement existing technologies to reduce their carbon footprint, the cost of such an endeavor will inevitably decrease. But if early adopters are helped and rewarded for their efforts, the rate of expansion of the NZEB concept will become

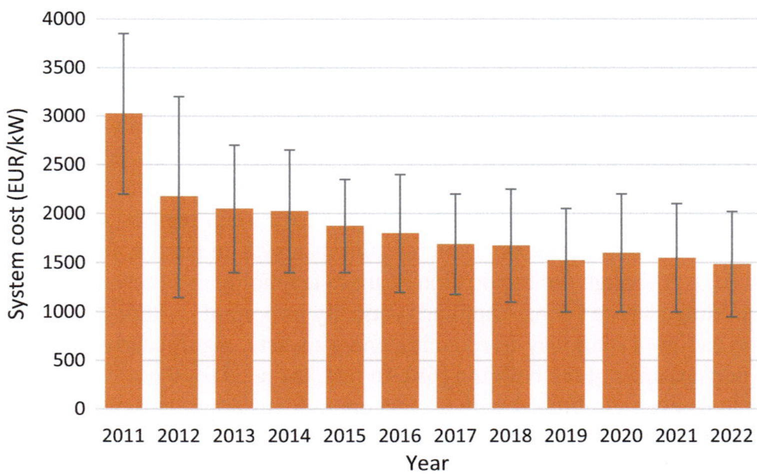

Fig. 8.3 Photovoltaic system prices in Germany from 2011 to 2022

Table 8.4 HVAC technology implementation

Technology type	Implementation issues	References
Photovoltaic (PV) panels	Most widely applied renewable energy source, 942GW capacity in 2021, 842% increase from 2011	[71]
Thermophotovoltaic panels	Because of its complexity, much development is still needed, although additional possible improvements of up to 32% have been proven	[72]
Concentrated solar power	Proven to be economically viable even with the considerable upfront cost when compared to PVs, allowing for higher solar power penetration	[73]
Small wind turbines	Considering the fluctuations in wind power availability, SWTs are shown to work best when combined with other energy sources and/or energy storage	[74]
Geothermal heat pumps	One of the most suitable technologies for renewable energy due to its high-performance factor and low-temperature heat energy	[75]
Biomass and biogas systems	The cost is considerably impacted by the cost of basic commodities	[76]
High efficiency HVAC	With optimization using modern methods, energy consumption could be reduced by over 5%	[77]
Energy storage	Storage capacity will have an average annual growth rate of 27% throughout this decade	[26]
Combined heat and power	Compared to central power stations and boilers, CHP can use up to 35% less primary energy	[27]
Solar thermal collector	Significant improvements can be made by using biodegradable instead of inorganic materials	[78]

considerably faster. Considering the urgency of the issue that is at hand, all factors ought to be contemplated, and any issues in the process of applying net-zero technologies must be solved. In countries or cities where there are incentives from the government for the creation of NZEBs, these technologies become more attractive and affordable. The European Union has been a global leader in incentivizing and prescribing as mandatory the move away from non-sustainable construction. In 2010 the European Parliament put forth a directive in which it sought to "honor both its long-term commitment to maintain the global temperature rise below 2 °C, and its commitment to reduce, by 2020, overall greenhouse gas emissions by at least 20% below 1990 levels" [79]. By setting minimum requirements for the energy performance of buildings it is practically mandated that all new buildings are nearly ZEB. This sort of action by governing bodies is a crucial step in the global fight against climate change. Governments can also permit the use of renewable sources or technologies which were either unavailable or illegal before. Creating a regulatory framework that allows individuals and companies to better understand and utilize sustainable building technologies will allow for a quicker adoption of net-zero technology and considerable reduction in carbon emissions [80]. Although the EU has made progress in this regard over the last two decades, more must be done by all regulatory bodies to ensure our transition from harmful technologies and energy sources, and with the feasibility of increasing proven subsidies [81]. Regulations and government subsidies are the first steps in successfully implementing net-zero technology. NGOs, government and other relevant large organizations and startups should actively educate the public on this issue, enabling informed decision-making [82]. Though this is a global issue, the solutions will not always be applicable on a global scale. Regional variables will determine the appropriate approach to reducing emissions. Many developing countries are experiencing energy insecurity, often relying on non-renewable sources, so helping these countries by investing in renewable infrastructure and making net-zero systems available and affordable could not only resolve energy shortages but also help with a quicker development of these regions [83]. Another unavoidable topic when it comes to the implementation of sustainable technologies in buildings is the effect geopolitical events on ability to focus on sustainability. Two large events, the COVID-19 pandemic, and the war in Ukraine, have had an undeniable impact on the West's ability to allocate resources to mitigating climate change [84]. Much of the focus went from sustainability to dealing with these crises. It can be assumed that there will be similarly impactful events in the future, and preparations must be made for these setbacks.

8.5 Carbon Capture and Storage Technology

While all listed technologies are important a play a significant role in lowering global carbon emissions, none of them promise to include non-renewable sources in our carbon neutral future. Considering the amount of existing infrastructure made for extracting non-renewable energy, and the global reliance on these traditional energy sources, it is important to consider ways to reduce the negative impact by using these systems. As the name suggests, this system functions by capturing CO_2 before it enters the atmosphere, storing it in underground geological formations [85]. With current capture rates of 85% to 90% [86], this technology could allow for the usage of fossil fuels without the environmental impact present currently. This is a field of research with significant funds, partly since it could potentially prolong the fossil fuel industry which means that this industry has a vested interest in researching this technology further, therefore much of the development of CCS is ahead of us. Research has shown that the CO_2 capture rate could be increased to as much as over 99% [87]. This process works in three steps: first it separates CO_2 from other emissions where it is produced and keeps it from entering the atmosphere, then the CO_2 is transported, and finally it is permanently stored in a designated carbon storage area [88]. Figure 8.4 shows the typical process of a CCS system.

These systems have been applied and tested since the 90s, increasing to dozens of active projects across the globe by the late 2000s [89]. Although the costs can be daunting, it is important to consider that in a world where carbon taxes are becoming more widely applied, the cost of carbon capture not only becomes more acceptable, but it is often the only way for large polluters to stay operational, since CCS allows

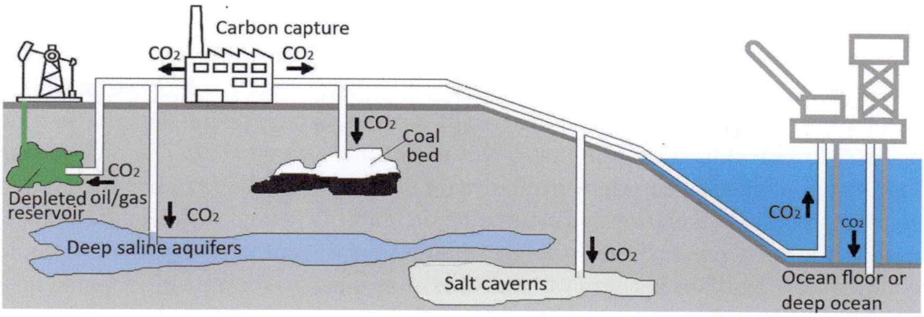

Fig. 8.4 Carbon capture and storage process

them to decrease or almost fully eliminate their carbon footprint [90]. As far as the storage goes, the main options currently available include storage in ocean floor, unmineable coal beds, depleted hydrocarbon reservoirs and deep saline aquifers, with saline aquifers having the highest capacity compared to the others [91]. Additionally, carbon capture systems can be divided into those who require significant modification of the original equipment and are often not compatible with existing infrastructure (pre-combustion and oxy-fuel-combustion) and those who can be easily retrofitted to existing infrastructure which allows for a far more cost-effective and approachable method of applying CCS systems (post-combustion) [92]. Another promising process which has proven quite cost-effective is the calcium looping cycle. Calcium looping is the reversible reaction between CaO and CO_2. It has proven highly promising with a small efficiency penalty (6–8%) and its cheap sorbent-crushed limestone [93].

8.6 Future Trends and Research Directions

The research and development of these technologies is still ongoing, and they are continually being improved on. Since the issue of climate change is one of society's highest priorities, efforts to reduce our carbon footprint have seen lots of funding and public support. With the current desire for increased sustainability, any technology that shows potential in its theoretical phase will presumably get enough attention to be further researched and developed, which means that the landscape of zero emission technology is ever changing. Among the topics that are most discussed related to carbon emissions has been the concept of carbon removal. Carbon removal takes carbon out of the atmosphere once it is introduced to it. In other words, it works retroactively, rather than exclusively where it is part of an ongoing energy production process. The promise of this concept is obvious, since it would allow us not only to reduce current emissions but fix some of the harm that has already been done to the environment. The main approaches include biological methods, which use forests, agricultural systems, and marine environments to capture and store carbon; geologic methods, which capture carbon dioxide by various means and store it underground in rock; and carbon-utilization methods, which capture carbon dioxide and use it to produce long-lived products such as plastics or cement [94]. One of the most researched types of carbon capture is direct air capture (DAC), which extracts CO_2 directly from the atmosphere, but the cost of the CO_2 captured, and the construction of the system is a large hurdle that is yet to be fully solved [95]. The unfortunate issue is that there has been an overreliance on this potential system when it comes to governing bodies calculating and estimating their ability to deal with climate change. In fact, if all pledges made regarding carbon removal were to be fulfilled, it would require over 1.2 billion hectares of land, almost as much as the area of cropland now [96]. Using renewable energy sources is crucial for a better future, but energy sources are not the only issue that must be addressed. For example, common materials used in construction such as paper, aluminum and steel are going to experience an even bigger demand while the supply will not be able to keep up. It is presumed that the supply will be 40% smaller than the demand [97]. This means that the reliance on these materials must be decreased in favor of zero-emission materials, whether they are natural, synthetic, or recycled. One consideration that is often left out of scientific research and calculations is the management of these systems. Using proper techniques, the efficiency of these technologies can be significantly improved. This is why more research, and development has been focused on creating smart energy management systems for buildings. By using *smart systems* instead of traditional ones, it can reduce the energy consumption of a building, since the energy usage will be dynamically set to always be at an ideal level unlike with manual input which inevitably wastes energy. A smart energy management system can immediately react to changes in energy supply, demand, weather changes and other constantly changing conditions [98]. Table 8.5 outlines key emerging trends that are expected to significantly influence the development and implementation of ZEBs in the coming years.

Referring to Table 8.5, it is observed that these trends and challenges highlight the dynamic and multifaceted approach needed to advance ZEBs. By integrating carbon removal technologies, adopting sustainable materials, and implementing smart energy systems and grids, the building sector can significantly reduce its environmental footprint. Continued innovation and strategic adoption of these technologies will be critical in achieving long-term sustainability.

8.7 Conclusion

Since buildings account for almost half of all energy consumption, it is crucial to focus on the ability to decrease the energy demand of buildings while simultaneously changing energy sources from non-renewable to renewable. By employing the usage of high-quality construction with eco-friendly materials, low waste and applying renewable energy sources such as solar and wind power, geothermal pumps, energy storage and others, it can ensure a future where buildings can easily have a net-zero impact on the environment. It is also crucial to change human behavior to where the

Table 8.5 Emerging future trends supporting ZEB and challenges

Future trends	Elaboration	Challenges	References
Carbon removal	Removes existing carbon from the atmosphere, using natural processes, storge underground, and utilizing carbon for creating various products	There is an overreliance on this technology since it is incredibly promising. Additionally, the costs of this system are still too great to easily implement on a large scale	[94]
Sustainable materials	By using sustainably sourced and produced materials and those that do not produce waste and considerable emissions, we can avoid overreliance on the most used materials which cannot keep up with the current demand	There are no easy replacements for the most commonly used construction materials, and in some cases it seems impossible to replace non-sustainable materials. Even in cases where traditional materials can be replaced by zero-emission materials, the supply of these materials is greatly lacking	[97]
Smart energy management systems	Using this system can allow for a significant improvement in the performance of zero-emission technologies applied in a building, allowing for the technologies to reach their true potential	Limited understanding of this system by potential users and building managers and a lack of interest of policymakers in promoting and subsidizing this technology make this system less popular than it easily could be considering its benefits	[99]
Smart grids	Allows for a much more reliable grid which has higher resilience to unexpected condition changes, while lowering costs by producing less energy waste. It can lower CO_2 emissions	As with any high-tech system, there is a vulnerability to malicious attacks on the grid which need to be detected and protected against to avoid grid failures. Additionally, a smart grid could potentially not be optimized for ideal performance as a lack of expertise by the grid manager	[100]

average person is educated on this topic and considers sustainable living paramount. Although the individual has a key role, much of the burden is on the shoulders of the politicians, policy makers, governments and organizations which can mandate a more sustainable practice in all aspects of life.

References

1. World Energy Outlook 2022. Accessed 01 Dec 2023. [Online]. Available: www.iea.org/t&c/
2. Jaysawal RK, Chakraborty S, Elangovan D, Padmanaban S (2022) Concept of net zero energy buildings (NZEB)—a literature review. Clean Eng Technol 11:100582. https://doi.org/10.1016/J.CLET.2022.100582

3. Tokimatsu K, Konishi S, Ishihara K, Tezuka T (2014) Global zero emission scenario: role of innovative technologies. Energy Procedia 61:164–167. https://doi.org/10.1016/J.EGYPRO.2014.11.930
4. Kilkis S (2009) A new metric for net-zero carbon buildings. Proc Energy Sustain Conf 2007:219–224. https://doi.org/10.1115/ES2007-36263
5. Dufo-López R, Bernal-Agustín JL (2012) Net zero energy buildings: expense or investment? Energy Procedia 14:1331–1336. https://doi.org/10.1016/J.EGYPRO.2011.12.1097
6. Turkenburg W et al (2012) Renewable energy. In: Johansson TB, Nakicenovic N, Patwardhan A, Gomez-Echeverri L (eds) Global energy assessment (GEA). Cambridge University Press, Cambridge, pp 761–900. https://doi.org/10.1017/CBO9780511793677.017
7. Tarazona-Romero BE, Socha-Rojas ND, Ascanio-Villabona J, Kafarov V, Castillo-Leon NY (2024) Performance evaluation of solar thermal collectors in Colombian thermal floors by dynamic simulation. Sustain Eng Innov 6(2):421–432. https://doi.org/10.37868/SEI.V6I2.ID394
8. Duraković B (2020) PCMs in separate heat storage modules. Green Energy Technol 121–146. https://doi.org/10.1007/978-3-030-38335-0_6/COVER
9. Summ T, Oyinlola M, Khattak S, Trinkl C, Zörner W (2024) Statistical analysis of solar thermal collectors in the Solar Keymark Database. Sustain Energy Technol Assess 61:103581. https://doi.org/10.1016/J.SETA.2023.103581
10. Khargotra R, Alam T, Thu K, Sebestyén V, András K, Singh T (2023) Experimental study of eco-friendly insulating materials for solar thermal collectors: a sustainable built environment. Results Eng 101681. https://doi.org/10.1016/J.RINENG.2023.101681
11. Sarhaddi F, Farahat S, Ajam H, Behzadmehr A (2010) Exergetic performance assessment of a solar photovoltaic thermal (PV/T) air collector. Energy Build 42(11):2184–2199. https://doi.org/10.1016/J.ENBUILD.2010.07.011
12. ALRubeei IRN, Idi SN, Hussein Alsammak IL, AlRikabi HT, Mutar HA, Alaidi AHM (2025) Using artificial intelligence for enhancement of solar cell efficiency in the of Iraq. Herit Sustain Dev 7(1):197–210. https://doi.org/10.37868/HSD.V7I1.1067
13. Hu A et al (2015) Impact of solar panels on global climate. Nature Climate Change 6(3):290–294. https://doi.org/10.1038/nclimate2843
14. Rabaia MKH et al (2021) Environmental impacts of solar energy systems: a review. Sci Total Environ 754. https://doi.org/10.1016/j.scitotenv.2020.141989
15. Benda V, Černá L (2020) PV cells and modules—state of the art, limits and trends. https://doi.org/10.1016/j.heliyon.2020.e05666
16. Lenert A et al (2014) A nanophotonic solar thermophotovoltaic device. Nat Nanotechnol 9(2):126–130. https://doi.org/10.1038/nnano.2013.286
17. Sovacool BK, Barnacle ML, Smith A, Brisbois MC (2022) Towards improved solar energy justice: exploring the complex inequities of household adoption of photovoltaic panels. Energy Policy 164:112868. https://doi.org/10.1016/J.ENPOL.2022.112868
18. Sodhi M, Banaszek L, Magee C, Rivero-Hudec M (2022) Economic lifetimes of solar panels. Procedia CIRP 105:782–787. https://doi.org/10.1016/J.PROCIR.2022.02.130
19. Iec, IEC 61400-2 INTERNATIONAL STANDARD NORME INTERNATIONALE Wind turbines-Part 2: Small wind turbines Eoliennes-Partie 2: Petits aérogénérateurs
20. Messineo A, Culotta S (2012) Evaluating the performances of small wind turbines: a case study in the South of Italy. Energy Procedia 16:137–145. https://doi.org/10.1016/J.EGYPRO.2012.01.024
21. Zajicek L, Drapalik M, Kral I, Liebert W (2023) Energy efficiency and environmental impacts of horizontal small wind turbines in Austria. Sustain Energy Technol Assess 59:103411. https://doi.org/10.1016/J.SETA.2023.103411
22. Rosato A, Perrotta A, Maffei L (2024) Commercial small-scale horizontal and vertical wind turbines: a comprehensive review of geometry, materials, costs and performance. Energies 17(13):3125. https://doi.org/10.3390/EN17133125
23. The efficiency of horizontal and vertical windmills: a comparative study—Naier. Accessed 16 May 2025. [Online]. Available: https://www.smartwindturbine.com/a-the-efficiency-of-horizontal-and-vertical-windmills-a-comparative-study.html?utm_source=chatgpt.com

24. Small wind electric systems a guide produced for the tennessee valley authority energy efficiency and renewable energy small wind electric systems
25. Al-Rawajfeh MA, Gomaa MR (2023) Comparison between horizontal and vertical axis wind turbine. Int J Appl Power Eng (IJAPE) 12(1):13–23. https://doi.org/10.11591/ijape.v12.i1.pp13-23
26. Hall CAS, Lambert JG, Balogh SB (2014) EROI of different fuels and the implications for society. Energy Policy 64:141–152. https://doi.org/10.1016/j.enpol.2013.05.049
27. Enevoldsen P, Xydis G (2019) Examining the trends of 35 years growth of key wind turbine components. https://doi.org/10.1016/j.esd.2019.02.003
28. McKenna R et al (2022) High-resolution large-scale onshore wind energy assessments: a review of potential definitions, methodologies and future research needs. https://doi.org/10.1016/j.renene.2021.10.027
29. Guo T et al (2021) Nacelle and tower effect on a stand-along wind turbine energy output—a discussion on field measurements of a small wind turbine. Appl Energy 303. https://doi.org/10.1016/j.apenergy.2021.117590
30. Moreira Chagas CC, Pereira MG, Rosa LP, da Silva NF, Vasconcelos Freitas MA, Hunt JD (2020) From megawatts to kilowatts: a review of small wind turbine applications, lessons from the US to Brazil. https://doi.org/10.3390/su12072760
31. Amjith LR, Bavanish B (2022) A review on biomass and wind as renewable energy for sustainable environment. Chemosphere 293. https://doi.org/10.1016/j.chemosphere.2022.133579
32. Xu Q et al (2023) A guided-liquid-based hybrid triboelectric nanogenerator for omnidirectional and high-performance ocean wave energy harvesting. Nano Energy 109. https://doi.org/10.1016/j.nanoen.2023.108240
33. Castillo-Leon NY, Rincon-Quintero AD, Mosquera-Clavijo AA, Suarez-Lopez CA, Betancur-Arboleda LA, del Portillo-Valdes L (2024) Energy and exergy evaluation in a densified biomass burner using cocoa shells, intended for air heating for the artificial drying of food. Sustain Eng Innov 6(2):309–326. https://doi.org/10.37868/SEI.V6I2.ID385
34. Wang K, Tester JW (2023) Sustainable management of unavoidable biomass wastes. Green Energy Resour 1(1):100005. https://doi.org/10.1016/J.GERR.2023.100005
35. Afridi ZUR et al (2023) Biogas as sustainable approach for social uplift in South East Asian Region. Energy Rep 10:4808–4818. https://doi.org/10.1016/J.EGYR.2023.11.037
36. Singh AK et al (2023) Sustainable utilization of biowaste resources for biogas production to meet rural bioenergy requirements. Energies 16(14):5409. https://doi.org/10.3390/EN16145409
37. Kabeyi MJB, Olanrewaju OA, Akpan J, Kabeyi MJB, Olanrewaju OA, Akpan J (2024) Biogas production and process control improvements. From Biomass Biobased Prod. https://doi.org/10.5772/INTECHOPEN.113061
38. Biogas from agricultural waste-a techno-economic evaluation final report. Accessed 16 May 2025. [Online]. Available: https://www.racefor2030.com.au
39. Kim D, Kim J (2025) Predicting biogas production from organic waste through anaerobic co-digestion. J Clean Prod 496:145122. https://doi.org/10.1016/J.JCLEPRO.2025.145122
40. Tagne RFT, Dong X, Anagho SG, Kaiser S, Ulgiati S (2021) Technologies, challenges and perspectives of biogas production within an agricultural context. The case of China and Africa. Environ Dev Sustain 23(10):14799–14826. https://doi.org/10.1007/S10668-021-01272-9/FIGURES/9
41. Tshikovhi A, Motaung TE (2023) Technologies and innovations for biomass energy production. Sustainability 15(16):12121. https://doi.org/10.3390/SU151612121
42. Sher F et al (2025) Cutting-edge biomass gasification technologies for renewable energy generation and achieving net zero emissions. Energy Convers Manag 323:119213. https://doi.org/10.1016/J.ENCONMAN.2024.119213
43. Sivabalan K, Hassan S, Ya H, Pasupuleti J (2021) A review on the characteristic of biomass and classification of bioenergy through direct combustion and gasification as an alternative power supply. J Phys Conf Ser. https://doi.org/10.1088/1742-6596/1831/1/012033

44. Wang Y, Liu Y, Dou J, Li M, Zeng M (2020) Geothermal energy in China: status, challenges, and policy recommendations. Util Policy 64. https://doi.org/10.1016/j.jup.2020.101020
45. van der Zwaan B, Dalla Longa F (2019) Integrated assessment projections for global geothermal energy use. Geothermics 82. https://doi.org/10.1016/j.geothermics.2019.06.008
46. Gibbons L, Javed S (2022) A review of HVAC solution-sets and energy performace of nearly zero-energy multi-story apartment buildings in Nordic climates by statistical analysis of environmental performance certificates and literature review. Energy 238:121709. https://doi.org/10.1016/J.ENERGY.2021.121709
47. Marigo M, Maccarini A, Zarrella A, Afshari A (2023) Fault impact analysis of ventilation systems in residential buildings: a simulation-based case study in Denmark. Energy Build 292:113150. https://doi.org/10.1016/J.ENBUILD.2023.113150
48. Nassif N, Moujaes S (2008) A cost-effective operating strategy to reduce energy consumption in a HVAC system. Int J Energy Res 32(6):543–558. https://doi.org/10.1002/ER.1364
49. Ghoreishi-Madiseh SA, Kuyuk AF (2017) A techno-economic model for application of geothermal heat pump systems. Energy Procedia 142:2611–2616. https://doi.org/10.1016/J.EGYPRO.2017.12.200
50. O'Hegarty R, Kinnane O, Lennon D, Colclough S (2022) Air-to-water heat pumps: review and analysis of the performance gap between in-use and product rated performance. Renew Sustain Energy Rev 155:111887. https://doi.org/10.1016/J.RSER.2021.111887
51. Pospíšil J, Špiláček M, Kudela L (2018) Potential of predictive control for improvement of seasonal coefficient of performance of air source heat pump in Central European climate zone. Energy 154:415–423. https://doi.org/10.1016/J.ENERGY.2018.04.131
52. Adebayo P et al (2024) Development, modeling, and optimization of ground source heat pump systems for cold climates: a comprehensive review. Energy Build 320:114646. https://doi.org/10.1016/J.ENBUILD.2024.114646
53. Valancius R, Singh RM, Jurelionis A, Vaiciunas J (2019) A review of heat pump systems and applications in cold climates: evidence from Lithuania. Energies 12(22):4331. https://doi.org/10.3390/EN12224331
54. Li J, Xiao Y, Lu S (2024) Optimal configuration of multi microgrid electric hydrogen hybrid energy storage capacity based on distributed robustness. J Energy Storage 76:109762. https://doi.org/10.1016/J.EST.2023.109762
55. Duraković B, Halilović M, Ali HM (2023) Phase change materials applications in buildings. In: Phase change materials for heat transfer, pp 225–248. https://doi.org/10.1016/B978-0-323-91905-0.00005-8
56. Taye BA, Choudhury NBD (2023) Adaptive filter based method for hybrid energy storage system management in DC microgrid. e-Prime—Adv Electr Eng Electron Energy 5:100259. https://doi.org/10.1016/J.PRIME.2023.100259
57. Migliari L, Petrollese M, Cau G, Cocco D (2024) Techno-economic assessment and grid impact of Thermally-Integrated Pumped Thermal Energy Storage (TI-PTES) systems coupled with photovoltaic plants for small-scale applications. J Energy Storage 77:109898. https://doi.org/10.1016/J.EST.2023.109898
58. 2H 2023 energy storage market outlook|BloombergNEF. Accessed 17 Dec 2023. [Online]. Available: https://about.bnef.com/blog/2h-2023-energy-storage-market-outlook/
59. Mundada AS, Shah KK, Pearce JM (2016) Levelized cost of electricity for solar photovoltaic, battery and cogen hybrid systems. Renew Sustain Energy Rev 57:692–703. https://doi.org/10.1016/J.RSER.2015.12.084
60. Garde F, Ayoub J, Aelenei D, Aelenei L, Scognamiglio A, Ernst W, Sohn (2017) Solution sets for net-zero energy buildings feedback from 30 buildings worldwide. Solut Sets Net-Zero Energy Build (March). https://doi.org/10.1002/9783433604663
61. Pearce JM (2009) Expanding photovoltaic penetration with residential distributed generation from hybrid solar photovoltaic and combined heat and power systems. Energy 34(11):1947–1954. https://doi.org/10.1016/J.ENERGY.2009.08.012
62. Duraković B, Hadziabdić M, Buyukdagli O (2023) Building energy demand management strategies and methods. In: Building energy flexibility and demand management. Academic Press, pp 63–85. https://doi.org/10.1016/B978-0-323-99588-7.00007-9

63. Tan Z et al (2019) Robust scheduling optimization model for multi-energy interdependent system based on energy storage technology and ground-source heat pump. Processes 7(1). https://doi.org/10.3390/PR7010027
64. Li P et al (2021) Research on thermo-economic characteristics of a combined cooling, heating and power system based on advanced adiabatic compressed air energy storage. J Energy Storage 103590. https://doi.org/10.1016/J.EST.2021.103590
65. Midilli A, Dogru M, Howarth CR, Ling MJ, Ayhan T (2001) Combustible gas production from sewage sludge with a downdraft gasifier. Energy Convers Manag 42(2):157–172. https://doi.org/10.1016/S0196-8904(00)00053-4
66. Le ST, Nguyen TN, Bui DK, Teodosio B, Ngo TD (2023) Comparative life cycle assessment of renewable energy storage systems for net-zero buildings with varying self-sufficient ratios. Energy 130041. https://doi.org/10.1016/J.ENERGY.2023.130041
67. Mavrigiannaki A et al (2021) Examining the benefits and barriers for the implementation of net zero energy settlements. Energy Build 230:110564. https://doi.org/10.1016/J.ENBUILD.2020.110564
68. Piccardo C, Dodoo A, Gustavsson L (2020) Retrofitting a building to passive house level: a life cycle carbon balance. Energy Build 223:110135. https://doi.org/10.1016/J.ENBUILD.2020.110135
69. Truong H, Garvie AM (2017) Chifley passive house: a case study in energy efficiency and comfort. Energy Procedia 121:214–221. https://doi.org/10.1016/J.EGYPRO.2017.08.020
70. Kraschewski T, Brauner T, Heumann M, Breitner MH (2023) Disentangle the price dispersion of residential solar photovoltaic systems: evidence from Germany. Energy Econ 121:106649. https://doi.org/10.1016/J.ENECO.2023.106649
71. Yolcan OO (2023) World energy outlook and state of renewable energy: 10-year evaluation. Innov Green Dev 2(4):100070. https://doi.org/10.1016/J.IGD.2023.100070
72. Khairul Azri AA et al (2023) Advancement in thermophotovoltaic technology and nanoparticle incorporation for power generation. Solar Energy 259:279–297. https://doi.org/10.1016/J.SOLENER.2023.05.018
73. Miron D, Navon A, Levron Y, Belikov J, Rotschild C (2023) The cost-competitiveness of concentrated solar power with thermal energy storage in power systems with high solar penetration levels. J Energy Storage 72:108464. https://doi.org/10.1016/J.EST.2023.108464
74. Barzegar-Kalashani M, Seyedmahmoudian M, Mekhilef S, Stojcevski A, Horan B (2023) Small-scale wind turbine control in high-speed wind conditions: a review. Sustain Energy Technol Assess 60:103577. https://doi.org/10.1016/J.SETA.2023.103577
75. Zhao S et al (2023) Competitive study of a geothermal heat pump equipped with an intermediate economizer for various ORC working fluids. Case Stud Therm Eng 45:102954. https://doi.org/10.1016/J.CSITE.2023.102954
76. Králík T et al (2023) Biomass price as a key factor for the further development of biogas and biomethane use—methodology and policy implications. Sustain Energy Technol Assess 60:103492. https://doi.org/10.1016/J.SETA.2023.103492
77. Borja-Conde JA, Nadales JM, Ordonez JG, Fele F, Limon D (2024) Efficient management of HVAC systems through coordinated operation of parallel chiller units: an economic predictive control approach. Energy Build 304:113879. https://doi.org/10.1016/J.ENBUILD.2023.113879
78. Khargotra R, Alam T, Thu K, Sebestyén V, András K, Singh T (2024) Experimental study of eco-friendly insulating materials for solar thermal collectors: a sustainable built environment. Results Eng 21:101681. https://doi.org/10.1016/J.RINENG.2023.101681
79. Directive 2010/31/EU of the European Parliament and of the Council of 19 May 2010 on the energy performance of buildings
80. Xu D, Abbas S, Rafique K, Ali N (2023) The race to net-zero emissions: can green technological innovation and environmental regulation be the potential pathway to net-zero emissions? Technol Soc 75:102364. https://doi.org/10.1016/J.TECHSOC.2023.102364
81. Lee K (2023) Bold fiscal policies for a net-zero EU: promotion of electric vehicles and expansion of green energy. Energy Rep 10:2944–2949. https://doi.org/10.1016/J.EGYR.2023.09.138

82. Sindhwani R, Singh PL, Behl A, Afridi MS, Sammanit D, Tiwari AK (2022) Modeling the critical success factors of implementing net zero emission (NZE) and promoting resilience and social value creation. Technol Forecast Soc Change 181:121759. https://doi.org/10.1016/J.TECHFORE.2022.121759
83. Shakya SR et al (2023) Environmental, energy security, and energy equity (3E) benefits of net-zero emission strategy in a developing country: a case study of Nepal. Energy Rep 9:2359–2371. https://doi.org/10.1016/J.EGYR.2023.01.055
84. da Costa JP, Silva AL, Barcelò D, Rocha-Santos T, Duarte A (2023) Threats to sustainability in face of post-pandemic scenarios and the war in Ukraine. Sci Total Environ 892:164509. https://doi.org/10.1016/J.SCITOTENV.2023.164509
85. Jacobson MZ (2009) Review of solutions to global warming, air pollution, and energy security. Energy Environ Sci 2(2):148–173. https://doi.org/10.1039/B809990C
86. Feron P et al (2019) Towards zero emissions from fossil fuel power stations. Int J Greenhouse Gas Control 87:188–202. https://doi.org/10.1016/J.IJGGC.2019.05.018
87. Soepyan FB et al (2024) Optimization of a natural gas power plant with membrane and solid sorbent carbon capture systems. Carbon Capture Sci Technol 10:100165. https://doi.org/10.1016/J.CCST.2023.100165
88. Li JR et al (2011) Carbon dioxide capture-related gas adsorption and separation in metal-organic frameworks. Coord Chem Rev 255(15–16):1791–1823. https://doi.org/10.1016/J.CCR.2011.02.012
89. Michael K et al (2010) Geological storage of CO_2 in saline aquifers—a review of the experience from existing storage operations. Int J Greenhouse Gas Control 4(4):659–667. https://doi.org/10.1016/J.IJGGC.2009.12.011
90. Park S, Mun H, Park J, Lee I (2024) Cost optimization methodology based on carbon-techno-economic analysis: an application to post-combustion carbon capture process. J Clean Prod 434:139887. https://doi.org/10.1016/J.JCLEPRO.2023.139887
91. Ketzer JM, Iglesias RS, Einloft S (2015) Reducing greenhouse gas emissions with CO_2 capture and geological storage. In: Handbook of climate change mitigation and adaptation, pp 1–40. https://doi.org/10.1007/978-1-4614-6431-0_37-2
92. Zhang P, Yin P, Yang L, Cui X, Xing H, Suo X (2024) Recent advances and challenges in ionic materials for post-combustion carbon capture. Carbon Capture Sci Technol 11:100180. https://doi.org/10.1016/J.CCST.2023.100180
93. Blamey J, Anthony EJ, Wang J, Fennell PS (2010) The calcium looping cycle for large-scale CO_2 capture. Prog Energy Combust Sci 36(2):260–279. https://doi.org/10.1016/J.PECS.2009.10.001
94. Burns W (2018) Why talk about carbon removal? EcoRN: Clim Change Ecol (Topic). https://doi.org/10.17606/M6H66H
95. Okonkwo EC, AlNouss A, Shahbaz M, Al-Ansari T (2023) Developing integrated direct air capture and bioenergy with carbon capture and storage systems: progress towards 2 ° and 1.5 °C climate goals. Energy Convers Manag 296:117687. https://doi.org/10.1016/J.ENCONMAN.2023.117687
96. Perkins O, Alexander P, Arneth A, Brown C, Millington JDA, Rounsevell M (2022) Toward quantification of the feasible potential of land-based carbon dioxide removal. One Earth 6(12):1638–1651. https://doi.org/10.1016/J.ONEEAR.2023.11.011
97. Gast L, Allwood JM (2023) What bulk material production is possible on a transition to net zero emissions by 2050 with limited zero emissions resources? J Clean Prod 423:138346. https://doi.org/10.1016/J.JCLEPRO.2023.138346
98. Jia S, Sheng K, Huang D, Hu K, Xu Y, Yan C (2023) Design optimization of energy systems for zero energy buildings based on grid-friendly interaction with smart grid. Energy 284:129298. https://doi.org/10.1016/J.ENERGY.2023.129298
99. Rocha P, Siddiqui A, Stadler M (2015) Improving energy efficiency via smart building energy management systems: a comparison with policy measures. Energy Build 88:203–213. https://doi.org/10.1016/J.ENBUILD.2014.11.077

100. Lin H, Wang W, Zou Y, Chen H (2023) An evaluation model for smart grids in support of smart cities based on the hierarchy of needs theory. Glob Energy Interconnect 6(5):634–644. https://doi.org/10.1016/J.GLOEI.2023.10.009

Chapter 9
Building Automation System

Abstract Building automation system is enabler technology for zero energy buildings that ensures optimum energy performance, occupant comfort, and reliability. The automation system integrates all building subsystems including HVAC, lighting, energy management, and communication networks into a unified control system that enables real-time monitoring and decision-making. The system participates in demand response programs by adjusting building loads in response to grid signals contributing to operational costs reduction. Despite these benefits, certain challenges are identified related to standardization, interoperability, cybersecurity, and the large-scale deployment.

Keywords Building Automation systems · Renewable energy · HVAC · Energy storage · Passive design strategies · PV · Energy efficiency

9.1 Introduction

The growing challenges of resource depletion and climate change have made it crucial to shift towards an energy transition that focuses on reducing primary energy demand, improving energy efficiency, and promoting the widespread use of renewable energy sources. As part of this broader energy transition, specific measures are being implemented to address energy consumption in the building sector. To reduce the energy demand in buildings, the Directive 2010/31/EU on the energy performance of buildings mandates that all new construction and the stock of existing buildings be almost zero energy-efficient within a decade [1]. A Zero Energy Building (ZEB) is a structure with outstanding energy performance, where a significant portion of its energy demand is supplied by locally sourced or on-site renewable energy generation [2].

A building automation system (BAS) is a computer-based control system installed in buildings to monitor, regulate, and control the mechanical and electrical systems such as HVAC, lighting, power, and security systems. Therefore, a BAS is a subset of a building management system (BMS) focused on automating specific building

B. Duraković, *Zero Energy Buildings*, Green Energy and Technology,
https://doi.org/10.1007/978-3-032-16027-0_9

systems, while the BMS monitors, manages, and reports across multiple building systems.

BAS is crucial for regulating energy use in ZEB through complex, automated controls [3, 4]. BAS technology optimizes the interior climate of buildings with high operational requirements by continuously monitoring and adjusting energy demand and contributing to energy savings [4–6]. Figure 9.1 represents a typical classification of building automation system based on functionality.

Referring to Fig. 9.1, it is observed that BAS integrates various subsystems that have a specific functional area and focuses on particular building needs. These subsystems are briefly discussed in Table 9.1, where each functional area is outlined, their descriptions, and the key components associated with each of them.

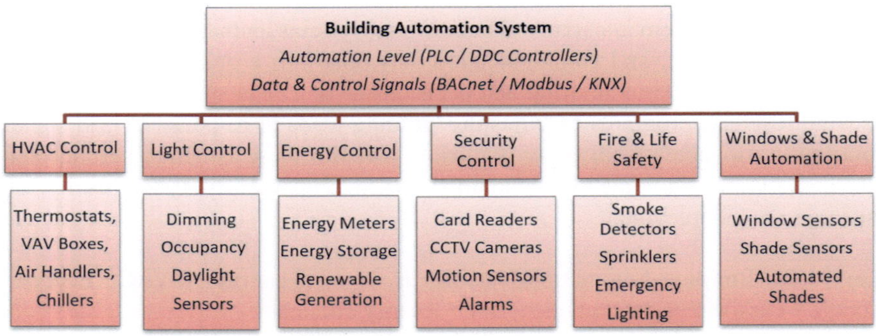

Fig. 9.1 Classification of building automation system based on functionality (Author: based on [7–9])

Table 9.1 BAS classification based on functionality

Functional area	Description	Key components
HVAC control	Controls temperature, ventilation, and air quality	Thermostats, AHUs, fans, dampers
Lighting control	Manages lighting levels and occupancy-based control	Occupancy sensors, daylight sensors
Security and access control	Ensures building security and restricts access	CCTV, access control, intrusion detection
Fire and life safety	Manages fire alarms, sprinklers, and emergency protocols	Smoke detectors, sprinkler systems
Energy management	Monitors and reduces energy consumption	Energy meters, smart controllers
Window and shade automation	Controls blinds, windows, and shades to optimize comfort	Motorized blinds, sunlight sensors
Communication and data integration	Enables communication and control across subsystems	Fieldbus systems, cloud integration

In Table 9.1, key BAS functional areas within a building are briefly summarized and discussed below. These systems work together seamlessly, enabling better control, monitoring, and maintenance of building operations, contributing to overall sustainability and operational effectiveness [7].

It is noteworthy that the biggest percentage of energy consumption in residential buildings is attributed to HVAC systems [10]. Air-source heat pumps (ASHP) are regarded as one of the energy-efficient HVAC devices that can raise the coefficient of performance (COP) of HVAC systems [11]. Moreover, different devices inside residential ZEBs are recognized as dispatchable resources in recognition of the importance of controlled loads in regulating energy usage [12–14].

Since *lighting* accounts for about 30% of building loads, it is especially important to include this load in building energy reduction efforts [15]. Energy savings from LED lighting and lower cooling/heating loads can reach $\leq 9\%$, making appropriate lighting equipment selection and investments some of the most economical ways to reduce CO_2 emissions [16]. CO_2 emissions reductions based on electrical lighting specifications are indicated 56–62% energy savings through improved lighting efficiency and decreased usage at night [17].

In general, BAS plays an important role in maximizing the performance of building energy storage, by monitoring variables such as state of charge (SoC), state of health (SoH), cell protection, real-time data, and fault detection for dependability. A previous study showed that low-power, wide-area network integrated with IoT ensures efficient real-time monitoring of the BAS of a Li-ion battery [18, 19]. For performance evaluation and condition analysis, a variation curve of the discharge of a battery is required to be monitored. In the applications of batteries, different kinds of loads can be connected, and each one affects the condition of the batteries in its own way. To ensure accurate measurement for analysis of a battery's SoC and SoH, some kinds of loads are necessarily connected [19–21].

The ability to monitor room occupancy and indoor air quality in real-time by an IoT sensor network allowed the acquisition of data such as room occupancy and indoor air quality [21]. Room occupancy was measured using occupant counters, while indoor pollutants such as carbon dioxide (CO_2), volatile organic compounds (VOC), can be measured with air quality sensors. Cloud computing technologies have enabled real-time data collection, processing, and storage and the simulation of comfort, energy, and health models and trade-off analysis in real time. All features could also be accessed in real time via an integrated web application thanks to the cloud-hosted platform [19].

Communication and data integration systems, including fieldbus networks and cloud integration, provide seamless communication and control across all subsystems, enabling centralized monitoring, remote control, and optimization. Together, these systems work in unison to improve the overall performance of the building, reduce operational costs, enhance energy efficiency, and create a safe and comfortable environment for its occupants [8, 9].

9.1.1 Automation and Control Systems for ZEB Performance

Automation and control systems are enabler components for ZEBs and improvement of tis performances [22]. Thanks to advanced technologies like predictive maintenance systems, it can early detect potential problems and cut energy losses [23]. These systems may use Model Predictive Control (MPC) approaches to calculate energy demands and make real-time adjustments to heating, cooling, and ventilation systems [24, 25]. Continuous monitoring and data collection is made through wireless sensor networks [26]. With the use of artificial intelligence (AI) and machine learning algorithms to automate complex decision-making processes [27], ZEBs can be adaptable to changing conditions and energy requirements. These technologies are enablers for the energy storage and renewable energy systems to operate continuously [28].

The implementation of thermal energy storage (TES) and battery energy storage systems (BESS) into ZEBs support the operation ensuring consistent performance under various load scenarios. Cold-water storage tanks or phase change materials may be used buffers between energy generation and building consumption [29]. BESS may be used for storing energy from photovoltaic (PV) systems used later when the demand is high or power prices are high [29, 30]. Automated control techniques, such as demand-response, guarantee that HVAC systems operate in a variety of scenarios. These covers may cover direct cooling (supplying energy straight from the chiller), discharging (using stored energy to satisfy cooling demands), and charging (storing energy for later use) [30]. Therefore, advanced control algorithms are required to adjust changing occupant needs, prioritizing cost reduction and energy efficiency [30, 31].

Various TES and BESS configurations are considered in the design and optimization of automation systems for ZEBs to maximize energy savings and reduce expenses. To provide energy flexibility and cost savings is demonstrated by the case study of an office building in Turin, Italy. The case showed how the automation-driven control of integrated TES and BESS with PV systems promote resilience and sustainability in building environments [30–32].

9.2 Demand Response and BAS

Demand Response (DR) is a strategy used to balance electricity supply and demand during peak periods when the grid is under stress. To adjust building energy consumption in real-time based on signals from utilities, BAS is crucial in facilitating this DR. During DR events, BAS automatically adjusts building subsystem and the other energy-consuming devices without compromising occupant comfort. This way, BAS supports grid stability and allows building owners to participate in DR programs [33, 34]. Table 9.2 summarizes comparison of DR and BAS and provides an overview of how these technologies work individually and together.

Table 9.2 BAS classification based on functionality

Aspect	Demand Response (DR)	Building Automation System (BAS)	Integration of DR and BAS
Definition	A strategy to reduce or shift energy usage during peak demand periods to balance supply and demand	A system that automates and optimizes building functions (HVAC, lighting, security) to improve energy efficiency and occupant comfort	Combining DR strategies with BAS to automatically manage energy use in response to grid signals
Primary goal	Reduce strain on the grid and lower energy costs	Improve energy efficiency, comfort, and building management	Optimize energy use during peak demand while ensuring minimal disruption to occupants
Key features	• Time-of-use pricing Incentive programs • Automated load control	• HVAC control • Lighting control • Energy monitoring • Security integration	• Automated energy reductions-Real-time data analysis • Peak load management
Benefits	• Lower energy costs • Increased grid stability • Environmental benefits (reduced emissions)	• Reduced operational costs • Improved comfort • Energy savings	• Increased energy savings-Reduced peak demand impact-Financial incentives from utilities
How it works	Consumers are incentivized to reduce energy usage during peak times	Centralized control and monitoring of building systems to improve energy efficiency	BAS automatically adjusts systems (HVAC, lighting) based on DR signals to reduce energy consumption during peak times
Challenges	• System integration issues • Consumer awareness	• Initial investment • System compatibility with DR programs	• Need for proper system integration • Potential communication failures between DR and BAS
Contribution to Sustainability	• Reduces need for additional power generation • Lowers carbon footprint	• Lowers building energy consumption • Helps meet sustainability goals	• Supports carbon reduction • Promotes energy • Efficient building operations

Referring to Table 9.2, it can be observed that DR focuses on adjusting energy consumption during peak demand periods, which helps balance the grid and reduce costs. Once these two are integrated, they work together to automate energy reductions as response to DR signals [35]. However, challenges such as system integration and consumer awareness need to be addressed for maximum effectiveness [36, 37]. On the other hand, the advancements in AI and IoT technologies enable BAS to predict energy demand patterns based on historical data and optimize responses to DR events. Smart thermostats application within the building enables real-time monitoring and control of HVAC systems.

Building energy planning under realistic scenario can be effectively managed by the HVAC system as the largest energy consumer in buildings [38, 39]. HVAC is a good candidate for DR strategies, and it gives some flexibility. HVAC can adjust its energy consumption by 40–65% during DR events [40]. AI-driven HVAC systems could reduce energy consumption by 15.8% annually in a commercial building. Additionally, if HVAC used renewable energy generation and energy storage, buildings could achieve near-zero energy consumption, contributing to sustainability goals and reducing reliance on the grid [41]. As buildings become smarter and more interconnected, the role of BAS in DR is expected to grow [42, 43].

9.3 Technological Emerging Trends and Challenges in BAS

BAS is an enabler technology that manages the energy used in buildings and optimizes its performance through embedded advanced technologies. The future trends are based on the integration of Artificial Intelligence (AI) and Machine Learning (ML) algorithms into BAS. These algorithms help to have predictive maintenance, energy optimization, and real-time data monitoring by analyzing data collected via Internet of Things (IoT) devices [42]. Additionally, the implementation of blockchain technology in BAS is gaining traction to address security and data integrity concerns [44].

AI was used in performance optimization energy storage systems, such as batteries or thermal storage, in ZEBs. By analyzing historical data AI can predict periods of high energy production or demand and adjust the storage and release of energy accordingly. This ensures that excess energy generated during peak renewable production times is efficiently stored and used during periods of low energy generation, reducing waste and maximizing the building energy autonomy.

Smart technologies facilitate the real-time oversight of renewable energy and energy storage systems, optimizing the operational modes of these storage systems, assessing their power capacity, and providing occupants with real-time advice regarding energy surpluses or deficits. At the connected edge devices of the IoT-based platform, valuable insights are derived from raw data through pre-processing [45]. An effective predictive maintenance framework for building systems includes five stages: gathering data; processing data; developing models; notifying of faults; enhancing the model to improve the predictive maintenance capabilities for the installation [45, 46].

Data gathered from diverse HVAC systems across various buildings, using IoT devices and a building automation system, serves as the input data. An IoT middleware solution was created to facilitate the connection and management of the IoT devices in these systems. The monitoring data from the installations undergoes a process involving data cleaning, organization, and feature engineering. A deep learning model is employed to forecast potential failures within the HVAC systems. A module for processing failure data is designed to handle information related to malfunctioning equipment [47]. The predictive maintenance model for installations is enhanced based on the outcomes from the identified malfunctions. A case study focusing on HVAC systems in a university building is included, demonstrating the effectiveness of the proposed framework in anticipating failures [45].

A growing number of zero-energy buildings that use renewable energy sources will be built, which will require storage systems. The developments in power electronics have made the use of energy storage systems practical. Consequently, there is a need for hybrid storage systems that can combine multiple renewable sources with energy storage ensuring a balanced and reliable energy supply. When there is a surplus of energy, storage systems become charged [45].

Combining a wind turbine with a battery storage system provides a stable and reliable power solution by balancing the wind's stochastic (unpredictable) nature and compensating for the delayed start-up of diesel generators, while also reducing costs and emissions [45].

9.4 Challenges in Building Automation Systems

Several challenges that hinder the widespread adoption of BAS are identified as interoperability, standardization and cybersecurity. Interoperability of equipment from multiple manufacturers which may utilize different communication protocols is a significant issue. So, the lack of standardization may complicate the integration of the system, underutilization of the generated data, as users may lack the tools or expertise to analyze and interpret them effectively [48, 49].

Cybersecurity of the BAS is another critical concern as the networks may have insufficient security measures that can lead to unauthorized access and manipulation of building systems. Implementing robust cybersecurity protocols and regular system audits is essential in mitigating these risks [50].

There is an understanding of the potential risk that arises from the fast rise of ZEBs if not properly planned. The rapid spread of ZEBs will impact residential and commercial energy demand and, indirectly, the network demand. The main concerns associated with the growing demand for the electric grid from new ZEBs include thermally driven and DC-coupled loads, greater weekday and seasonal variability, shifting peak demand patterns, an increased peak-to-average load ratio, and the need for effective resource allocation for energy storage and system balancing.

To reduce the perceived risk on the electricity network by the ZEB dissemination, it is possible to fine-tune the regulatory framework and present a set of policies and incentives to be adopted by governments to foster the ZEBs and energy storage systems uptake and implementation [45, 51]. Addressing these challenges is crucial for realizing the full potential of BAS in creating efficient, secure, and sustainable building environments.

9.5 Conclusion

BAS is a fundamental enabler of the ZEBs making them responsive to the environment and occupant comfort. This system coordinates various building subsystems including HVAC, appliances, energy management, etc. enabling the building to operate intelligently by reducing its energy consumption and saving the costs. Advanced AI-based technologies enhance system adaptability to real-time needs by optimizing building operations, providing predictive maintenance and management of thermal and electrical storage systems. These advanced automation strategies incorporated with energy storage systems and renewable energy, provide autonomy and resilience to buildings. However, several challenges remain that are related to interoperability, standardized communication protocols, vulnerabilities, and widespread adoption, which require development of new regulations, standards and technological innovation.

References

1. Directive 2010/31/EU of the European Parliament and of the Council of 19 May 2010 on the energy performance of buildings
2. Iturriaga E, Aldasoro U, Campos-Celador A, Sala JM (2017) A general model for the optimization of energy supply systems of buildings. Energy 138:954–966. https://doi.org/10.1016/J.ENERGY.2017.07.094
3. Siasiev A, Dudnikov S, Nikolaienko D, Hubal H, Bondarchuk O (2025) Evaluation of the effectiveness of decision-making algorithms in the management of renewable energy sources: a case study of Ukraine. Sustain Eng Innov 7(1):87–96. https://doi.org/10.37868/SEI.V7I1.ID415
4. Mariano-Hernández D, Hernández-Callejo L, Zorita-Lamadrid A, Duque-Pérez O, Santos García F (2021) A review of strategies for building energy management system: model predictive control, demand side management, optimization, and fault detect and diagnosis. J Build Eng 33:101692. https://doi.org/10.1016/J.JOBE.2020.101692
5. Levesque A, Pietzcker RC, Baumstark L, De Stercke S, Grübler A, Luderer G (2018) How much energy will buildings consume in 2100? A global perspective within a scenario framework. Energy 148:514–527. https://doi.org/10.1016/J.ENERGY.2018.01.139
6. Jain R et al (2023) Optimization of energy consumption in smart homes using firefly algorithm and deep neural networks. Sustain Eng Innov 5(2):161–176. https://doi.org/10.37868/SEI.V5I2.ID210

7. Dounis AI (2010) Artificial intelligence for energy conservation in buildings. Adv Build Energy Res 4(1):267–299. https://doi.org/10.3763/ABER.2009.0408;WGROUP:STRING:PUBLICATION
8. Bhatt J, Verma HK (2015) Design and development of wired building automation systems. Energy Build 103:396–413. https://doi.org/10.1016/J.ENBUILD.2015.02.054
9. van Roosmale S, Hellinckx P, Meysman J, Verbeke S, Audenaert A (2024) Building automation and control systems for office buildings: technical insights for effective facility management—a literature review. J Build Eng 97:110943. https://doi.org/10.1016/J.JOBE.2024.110943
10. Cetin KS, Fathollahzadeh MH, Kunwar N, Do H, Tabares-Velasco PC (2019) Development and validation of an HVAC on/off controller in EnergyPlus for energy simulation of residential and small commercial buildings. Energy Build 183:467–483. https://doi.org/10.1016/J.ENBUILD.2018.11.005
11. Wu W, Skye HM, Domanski PA (2018) Selecting HVAC systems to achieve comfortable and cost-effective residential net-zero energy buildings. Appl Energy 212:577–591. https://doi.org/10.1016/J.APENERGY.2017.12.046
12. Zhang J, Mu Y, Li J, Tong Y, Yang W (2024) Energy performance of a residential zero energy building energy system—R-CELLS at solar decathlon China 2022. Appl Energy 371:123742. https://doi.org/10.1016/J.APENERGY.2024.123742
13. Shaterabadi M, Mehrjerdi H, Amiri N, Ahmadi Jirdehi M, Iqbal A (2023) Green roof positive impact on changing a Plus-ZEB to drastic Plus-ZEB: the multi-objective energy planning and audit in real condition outlook. Energy Build 287:112983. https://doi.org/10.1016/J.ENBUILD.2023.112983
14. Gergely LZ, Barna E, Horváth M, Szalay Z (2025) Assessing embodied and operational carbon of residential HVAC systems: baselines for life-cycle sustainability. Build Environ 269:112442. https://doi.org/10.1016/J.BUILDENV.2024.112442
15. Khan N, Abas N (2011) Comparative study of energy saving light sources. Renew Sustain Energy Rev 15(1):296–309. https://doi.org/10.1016/J.RSER.2010.07.072
16. Ahn BL, Jang CY, Leigh SB, Yoo S, Jeong H (2014) Effect of LED lighting on the cooling and heating loads in office buildings. Appl Energy 113:1484–1489. https://doi.org/10.1016/J.APENERGY.2013.08.050
17. Lee R, Choi M, Yoon J, Kim D (2023) Impacts of lighting and plug load variations on residential building energy consumption targeting zero energy building goals. J Build Eng 75:106962. https://doi.org/10.1016/J.JOBE.2023.106962
18. Krishna G, Singh R, Gehlot A, Shaik VA, Twala B, Priyadarshi N (2024) IoT-based real-time analysis of battery management system with long range communication and FLoRa. Results Eng 23:102770. https://doi.org/10.1016/J.RINENG.2024.102770
19. Xu Y, Zhu S, Cai J, Chen J, Li S (2025) A large language model-based platform for real-time building monitoring and occupant interaction. J Build Eng 100:111488. https://doi.org/10.1016/J.JOBE.2024.111488
20. Nyamathulla S, Dhanamjayulu C (2024) A review of battery energy storage systems and advanced battery management system for different applications: challenges and recommendations. J Energy Storage 86:111179. https://doi.org/10.1016/J.EST.2024.111179
21. Selvaraj R, Kuthadi VM, Baskar S (2023) Smart building energy management and monitoring system based on artificial intelligence in smart city. Sustainable Energy Technol Assess 56:103090. https://doi.org/10.1016/J.SETA.2023.103090
22. Liu Z, Li M, Ji W (2025) Development and application of a digital twin model for Net zero energy building operation and maintenance utilizing BIM-IoT integration. Energy Build 328:115170. https://doi.org/10.1016/J.ENBUILD.2024.115170
23. Wang J et al (2024) Global assessment of radiative cooling and low-emissivity coatings for energy efficiency in buildings across all climate zones. Energy Build 317:114402. https://doi.org/10.1016/J.ENBUILD.2024.114402
24. Zhao J, Chen Z, Li H, Liu D (2024) A model predictive control for a multi-chiller system in data center considering whole system energy conservation. Energy Build 324:114919. https://doi.org/10.1016/J.ENBUILD.2024.114919

25. Stoffel P, Berktold M, Müller D (2024) Real-life data-driven model predictive control for building energy systems comparing different machine learning models. Energy Build 305:113895. https://doi.org/10.1016/J.ENBUILD.2024.113895
26. Ran J, Nan Y, Gao Q, Si P, Zhong H, Feng Y (2024) Beyond the g-value: a comparative study of solar control coated glass and spectrally selective glass in terms of building energy efficiency. Energy Build 325:114977. https://doi.org/10.1016/J.ENBUILD.2024.114977
27. Sánchez-García D, Martínez-Crespo J, Hernando URR, Alonso C (2023) A detailed view of the adaptive-comfort-control-implementation script (ACCIS): the capabilities of the automation system for adaptive setpoint temperatures in building energy models. Energy Build 288:113019. https://doi.org/10.1016/J.ENBUILD.2023.113019
28. Alhassan M, Alkhawaldeh A, Betoush N, Sawalha A, Amaireh L, Onaizi A (2024) Harmonizing smart technologies with building resilience and sustainable built environment systems. Results Eng 22. https://doi.org/10.1016/j.rineng.2024.102158
29. Brandi S, Gallo A, Capozzoli A (2022) A predictive and adaptive control strategy to optimize the management of integrated energy systems in buildings. Energy Rep 8:1550–1567. https://doi.org/10.1016/j.egyr.2021.12.058
30. Groumpos PP (2018) Advanced automation control systems (AACS) for energy and comfort management in a building environment. IFAC-PapersOnLine 34–38. https://doi.org/10.1016/j.ifacol.2018.11.241
31. Li H, Wang S (2020) Model-based multi-objective predictive scheduling and real-time optimal control of energy systems in zero/low energy buildings using a game theory approach. Autom Constr 113. https://doi.org/10.1016/j.autcon.2020.103139
32. Odufuwa OY, Kusakana K, Numbi BP, Tartibu LK Optimal energy management of grid-connected PV for HVAC cooling with ice thermal storage system. J Energy Storage 77. https://doi.org/10.1016/j.est.2023.109844
33. Jurjevic R, Zakula T (2023) Demand response in buildings: a comprehensive overview of current trends, approaches, and strategies. Buildings 2023 13(10):2663. https://doi.org/10.3390/BUILDINGS13102663
34. Khorram M, Abrishambaf O, Faria P, Vale Z (2018) Office building participation in demand response programs supported by intelligent lighting management. Energy Inform 1(1):1–14. https://doi.org/10.1186/S42162-018-0008-4/FIGURES/9
35. Duraković B, Hadziabdić M, Buyukdagli O (2023) Building energy demand management strategies and methods. In: Building energy flexibility and demand management, Academic Press, pp 63–85. https://doi.org/10.1016/B978-0-323-99588-7.00007-9
36. Drasch BJ, Fridgen G, Häfner L (2020) Demand response through automated air conditioning in commercial buildings—a data-driven approach. Bus Res 13(3):1491–1525. https://doi.org/10.1007/S40685-020-00122-0/TABLES/14
37. Ma Z, Jørgensen BN (2018) A discussion of building automation and stakeholder engagement for the readiness of energy flexible buildings. Energy Inform 1(1):1–15. https://doi.org/10.1186/S42162-018-0061-Z/FIGURES/2
38. Pallonetto F, De Rosa M, Milano F, Finn DP (2019) Demand response algorithms for smart-grid ready residential buildings using machine learning models. Appl Energy 239:1265–1282. https://doi.org/10.1016/J.APENERGY.2019.02.020
39. Wang H, Wang S, Tang R (2019) Development of grid-responsive buildings: opportunities, challenges, capabilities and applications of HVAC systems in non-residential buildings in providing ancillary services by fast demand responses to smart grids. Appl Energy 250:697–712. https://doi.org/10.1016/J.APENERGY.2019.04.159
40. Müller FL, Jansen B (2018) Large-scale demonstration of precise demand response provided by residential heating systems. Appl Energy 239:836–845. https://doi.org/10.1016/j.apenergy.2019.01.202
41. Hsu PC, Gao L, Hwang Y (2025) Comparative study of LSTM and ANN models for power consumption prediction of variable refrigerant flow (VRF) systems in buildings. Int J Refrig 169:55–68. https://doi.org/10.1016/j.ijrefrig.2024.10.020

42. Shah SFA et al (2022) The role of machine learning and the internet of things in smart buildings for energy efficiency. Appl Sci 12(15):7882. https://doi.org/10.3390/APP12157882
43. Demand Response and the Role of Building Automation in Grid Resilience—7NOX After Hours HVAC Scheduling. Accessed 21 May 2025 [Online]. Available: https://7nox.com/demand-response-and-the-role-of-building-automation-in-grid-resilience/?utm_source=chatgpt.com
44. Himeur Y, Ghanem K, Alsalemi A, Bensaali F, Amira A (2021) Artificial intelligence based anomaly detection of energy consumption in buildings: a review, current trends and new perspectives. Appl Energy 287. https://doi.org/10.1016/j.apenergy.2021.116601
45. Rahman MM, Oni AO, Gemechu E, Kumar A (2020) Assessment of energy storage technologies: a review. https://doi.org/10.1016/j.enconman.2020.113295
46. Yoon S, Song J, Li J (2025) Ontology-enabled AI agent-driven intelligent digital twins for building operations and maintenance. J Build Eng 108:112802. https://doi.org/10.1016/J.JOBE.2025.112802
47. Hulse P, Betancur-Arboleda L, Rincon-Quintero AD, Ascanio-Villabona JG, Tarazona-Romero BE (2024) FMECA and FTA analysis applied to the manufacturing process of pulsating heat pipes. Herit Sustain Dev 6(1):1–12. https://doi.org/10.37868/HSD.V6I1.319
48. Affonso EOT et al (2024) The main barriers limiting the development of smart buildings. Buildings 14(6):1726. https://doi.org/10.3390/BUILDINGS14061726/S1
49. van Roosmale S, Audenaert A, Meysman J (2024) Understanding the opportunities and challenges of building automation and control systems to support facility management—an extensive literature review. Facilities 42(7–8):677–693. https://doi.org/10.1108/F-05-2023-0042/FULL/XML
50. Li G et al (2023) A critical review of cyber-physical security for building automation systems. Annu Rev Control 55:237–254. https://doi.org/10.1016/j.arcontrol.2023.02.004
51. Adrian MM, Purnomo EP, Enrici A, Khairunnisa T (2023) Energy transition towards renewable energy in Indonesia. Herit Sustain Dev 5(1):107–118. https://doi.org/10.37868/HSD.V5I1.108

Chapter 10
Retrofitting Existing Buildings to Achieve Nearly Zero Energy Standards

Abstract Retrofitting existing residential and public buildings with energy-efficient measures has been investigated in this chapter. Particularly, in-depth case studies were analyzed and the result showed that retrofitting interventions, such as enhanced insulation, efficient glazing, and integration of renewable energy technologies lead to significant energy demand reduction and cost savings. In conclusion, this chapter could assist architects, engineers, and policymakers in implementing renewable energy in retrofitting existing building stock and establish energy-efficient building standards and policies.

Keywords Retrofitting · Energy efficiency · Sustainability · Building performance · Evolutionary algorithms

10.1 Introduction

Energy-efficient measures in residential buildings have become an important issue for researchers and contractors in developed countries due to the energy-saving trend that dominates their energy policy. The increasing concerns about climate change and sustainable living practices have led architects, engineers, and policymakers to investigate and find solutions for reducing energy consumption without compromising occupant comfort.

The energy transition and climate change pose a range of significant challenges to reduce energy demand in existing residential buildings. The pressure to implement energy conservation strategies has emerged as a critical challenge [1]. Energy use in public, commercial and residential buildings has risen significantly over the last two decades [2]. As a response, a variety of governmental efforts have been put into place, varying in intensity and structure, to increase energy efficiency in the construction sector [3].

The efficient use of energy in the built environment is a focal point on an international scale, especially in Europe, where retrofitting existing buildings is crucial for the reduction of energy consumption because of the low turnover rate of the

B. Duraković, *Zero Energy Buildings*, Green Energy and Technology,
https://doi.org/10.1007/978-3-032-16027-0_10

stock of buildings. While there is great potential for energy efficiency improvement, many opportunities are untapped because of insufficient methods for evaluating the effect of energy efficiency. In this chapter, some approaches to measure and verify energy savings, and predict and recommend energy retrofitting actions of buildings are discussed using data collected from the literature [4].

The purpose of this chapter is to investigate existing building retrofitting approaches and techniques that can contribute to lower heating and cooling energy demand in existing buildings. The transition to Net Zero Energy Buildings (NZEB) has become one of the important priorities of various governments across the world as they recognize the potential role that such structures can play in reducing the amount of energy consumed at a national level and in addressing climate change [5]. This chapter is based on a literature review with the objective of examining the primary parameters affecting building energy performance. The goal is to summarize current and emerging approaches and techniques employed in energy retrofitting, such as the use of energy-efficient materials, the implementation of advanced insulation techniques, high-efficiency heating, ventilation, and air conditioning (HVAC) systems, and the integration of renewable energy sources, particularly solar energy, into existing infrastructures.

10.2 Existing Approaches for Energy-Efficient Retrofitting

Existing retrofitting strategies encompass a range of approaches and techniques to reduce energy demand in existing buildings. Common approaches include upgrading building envelope insulation and use of high-performance windows, solar shading, improving HVAC efficiency, and integrating renewable energy technologies. Further, by adjusting design parameters such as envelope insulation thickness, window glazing type, heating and cooling set points, window-to-wall ratio, energy efficiency and cost-effectiveness can be optimized across different climatic zones [6].

The summary of the approaches for energy-efficient retrofitting is presented in Fig. 10.1. The figure illustrates the hierarchical relationship between strategies, approaches, and techniques in energy-efficient building retrofitting. Therefore, these retrofit solutions help lower energy bills, carbon footprints and extend the lifespan of buildings, contributing to more sustainable urban environments.

Referring to Fig. 10.1, a wide range of concrete actions and technologies that are applied on-site are listed at the bottom, including installing external insulation, fitting double/triple-glazed windows, installing solar shading devices [7], upgrading HVAC equipment, and installing solar panels. Overall, the figure provides a structured framework that explains how various retrofit efforts fit together to improve energy efficiency in existing buildings.

For instance, external envelope insulation, double-glazing windows, air tightness and solar shading were implemented in a hot arid climate resulting in 33% in energy reduction due to improved window glazing, and solar shading. The research further provides design codes to achieve thermal comfort and proposes a potential retrofitting

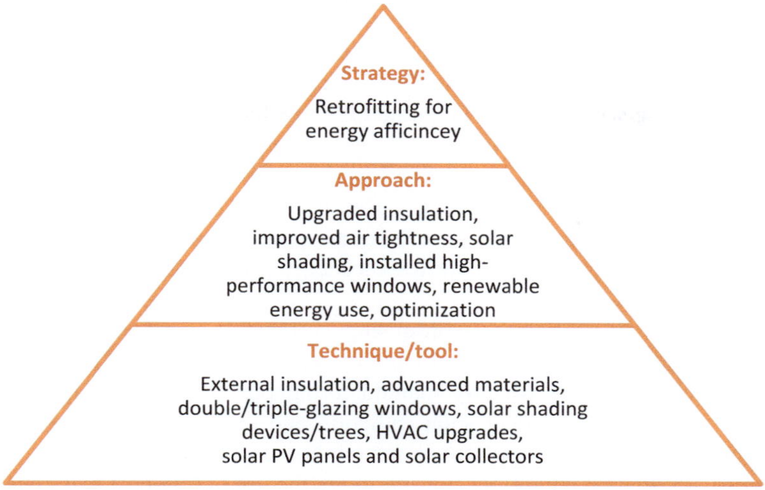

Fig. 10.1 Energy-efficient retrofitting approaches and techniques

strategy along with a reference to local energy efficiency baseline [7]. In addition, roof and exterior wall upgrades significantly reduced heat pump energy consumption by 33% in summer [8]. Air–water source heat pumps have the largest energy savings potential if building envelope was upgraded, than the window and door retrofits scored highest, followed by other elements of the building envelope. In terms of return on investment, roof retrofits were, on average, the most cost-effective where energy savings could be found [9].

Energy retrofit technology guarantees persistent and efficient energy consumption in modern educational buildings. To improve the diagnostic accuracy for exterior walls and windows, the performance assessment was performed with field measurement tools instead of numerical analysis. The energy performance of the target building was reported to be poor, implying a potential need for improvement measures aimed at better exploitation of energy. Efficiency and economic performance of three groups of retrofit technologies were studied including passive, active, and renewable solutions. The technologies that worked effectively were PV (photovoltaic) panels, HVAC (heating, ventilation, and air conditioning) systems, airtight films, and LED lighting. Based on energy savings, economic parameters, and thermal comfort, eight retrofit packages were developed. It was done by identifying an effective energy-saving package among packages like the energy-saving package, all-included technology package, economically beneficial package, and conservation package were identified. All we can say from the approach of the conservation package and the economically beneficial package, the study suggested them to be more economically efficient. The energy-saving package had a preferential effect on improving warm conditions and great improvements in cold thermal comfort. On the other hand, extra structural and aesthetic measures are needed to get a conservation package extractor that has energy-saving, economic, thermal comfort, and preservation advantages, which are all similar in contemporary aspects of a target building. The package proved to be the most economical solution for the target building, with a payback period of four years and a significant 41.4% energy savings [10].

The retrofitting with Building-Integrated Photovoltaics (BIPV) and Photovoltaic Thermal (PVT) systems enabled a renewable energy capacity of 75.4 kW and increased low-carbon energy production by 2.48 times, achieving a 46.2% energy surplus in the retrofitted community [11]. Retrofit approaches for achieving NZEBs emphasize the widespread use of photovoltaic panels across climates show that retrofitted residential NZEBs have the lowest energy consumption compared to other building types [5].

Passive radiative cooling is a promising sustainable solution for large-scale building cooling, with its commercial viability relaying to the use of low-cost materials and environmentally friendly manufacturing practices. The coating reflects approximately 97% of incident solar radiation and emits approximately 94% of the thermal radiation, which reaches a net cooling capacity of about 8 °C at high incident solar radiation conditions. Furthermore, the use of low-cost materials and a closed-loop process reduce the payback time and overall environmental impact by 54.96% and 28.29–54.19% with respect to commercial alternatives [12].

The integrated retrofitting of building heritage is of great interest in Europe, as shown by the number of projects being carried out every year in the framework of the H2020 research program that mainly focused on a new outer surface technology to improve both the seismic and energy performance of existing buildings [13].

10.2.1 Optimization Approaches

Optimization have crucial role in enhancing the energy efficiency of building retrofits by systematically balancing multiple objectives such as cost, energy savings, financial incentives, and payback periods. Most effective retrofit approaches within budget constraints, ensuring maximum return on investment can be selected. Algorithms such as multi-objective and heuristic optimization techniques can be applied in addressing these retrofit decisions. To improve decision-making in building retrofitting, different optimization approaches have been applied in the literature highlighting their potential such as Modified Supply–Demand-Based Optimization Algorithm (MSDOA), Genetic Algorithms (GAs), etc.

The MSDOA accuracy and precision was confirmed in nine case studies, with the aim of reducing the payback period and achieving energy savings. As an example, the project payback period was less than three years with an investment of $190,000 and energy savings were 10% of baseline consumption. The model includes constraints like initial investment, net present value, payback period, and energy. A sensitivity analysis was performed to test the sensitivity of the model under different assumptions, such as initial investments, potential misestimates of energy savings, auditing errors, fluctuations in electrical power costs, and interest rates. These results also show that more investment means more energy saving, but the payback period differs. The MSDOA converges with better results than other comparative algorithms as well as guarantees faster convergence, proving the effectiveness and advantage of MSDOA over other algorithms [14]. To optimize building retrofits GAs was applied in 175 case studies [15], due to a low computation cost and high-quality solutions. More current optimization strategies have been developed to include life-cycle analysis paired with dynamic simulation tools like EnergyPlus [15, 16].

The use of decision-making tools such as Cost–Benefit Analysis (CBA) for sustainable upgrades in existing buildings is a useful approach for demonstrating the economic evaluation of adopting new upgrades. A CBA framework integrates green building criteria to guide early-stage office renovation decisions, demonstrated through a case study in Maribor [17]. Dynamic-affordability and similar techniques are used most frequently and give a clear answer in terms of financial values, but they are affected by discount rates and assumptions. Although energy savings and greenhouse gas reductions are the primary concerns, other environmental and social benefits are often neglected, indicating the need for a more holistic assessment [18].

People generally have a positive perception of climate-sensitive factors and benefits, with 56% expressing favorable views due to increased comfort and energy savings [19]. Users are willing to pay significantly for co-benefits but also value equity and long-term environmental concerns in a way that indicates they would accept a degree of sacrifice of individual lifestyle [20].

Despite substantial advancements in optimization algorithms, several critical challenges persist, including long computational times, limited data availability, and the inadequate incorporation of occupant preferences and indoor environmental quality factors. Recognizing these limitations, and in order to create site-specific retrofit simulations using GA-based optimization a framework was established based on future needs in terms of accessibility new materials, tools, and technologies [21, 22]. The framework reliability and robustness are demonstrated through various case studies and prove cost-effective retrofit solutions [22, 23].

Based on the literature, GA-based multi-objective optimization is a good support for building retrofit optimization in achieving robust objective improvements. Challenges include high computational demands requiring hybrid techniques, limited focus on heritage buildings, lack of standardized methods, integration difficulties, and the expertise needed to manage analyses [24]. Despite these challenges, GA-based optimization has promising potential to support retrofit decision-making if combined with complementary approaches.

10.3 Building Envelope Upgrade

Case studies of a residential sector found that high energy use and greenhouse gas emissions are mainly due to insufficient insulation and out-of-date building techniques. Figure 10.2 shows the concept of energy-efficient retrofitting, a conversion of a conventional uninsulated building into its improved energy-efficient version.

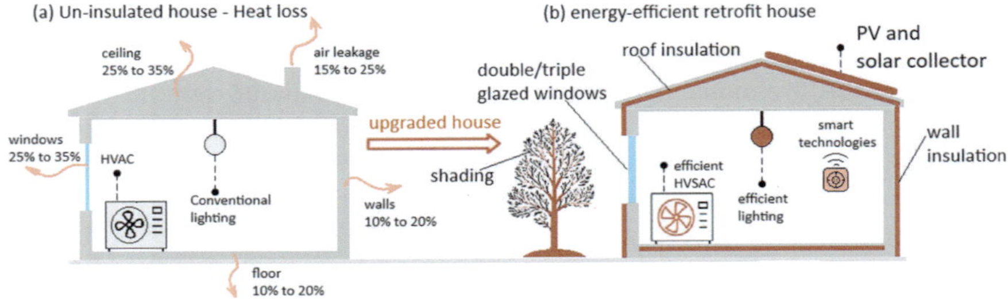

Fig. 10.2 Energy-efficient retrofitting: **a** traditional uninsulated building; **b** upgraded building for energy efficiency

Numerous older buildings do not conform to contemporary requirements in respect of thermal design, which call for effective retrofitting solutions. Recently proposed energy retrofit strategies leading to holistic approaches that improve building efficiency and reduce emissions. The limited use of sustainable materials in integrated systems suggests that further research and development is needed to advance resilient and energy-efficient construction [25]. Emerging building envelope material technologies such as Phase Change Material (PCM), aerogel, vacuum insulation panel, and heat reflective coating are often used as a retrofitting approach [26]. Even a certain technological progress has been made, there is still a gap in using advanced materials properly for building envelopes [24].

10.3.1 Insulation and Air Sealing

Residential buildings that were built before 1992 often suffer from excess air leakage and lack of insulation. Energy retrofits of these building can significantly improve energy performance and lower utility bills via moisture management and efficient building systems. The literature reviewed emphasizes the applicability of wall retrofit components to those in cold climates and generally focuses on low-cost, removable technologies that can be utilized over existing cladding [27].

Bearing in mind the degradation of external wall insulation over time is important for accurately assessing the energy-saving potential of retrofit projects in old buildings. The U-value of external walls must be measured and corrected using actual measurement data. The significant loss in energy-saving potential is evident when U-value deterioration is ignored. The insulation performance of the aged buildings can be decreased by 49.69% on average, relative to the original legal standard. If only legal standards were used to determine thermal transmittance, it was underestimated by 16.73% in terms of energy savings [28]. The unrenovated buildings are poorly designed for low-temperature heating and have difficulty maintaining comfort in the winter months. Therefore, it is necessary to apply insulation methods specific to the geographical area to ensure sufficient indoor comfort [29].

The application innovative materials in building envelopes such as PCM, bio-based concrete and hemp concrete play a vital role in maintaining indoor thermal comfort and energy consumption. While they each have their advantages, still there is limited research on their combined use and an in-depth evaluation of their energy and hygrothermal performance. Some available results demonstrate that placing PCM near the interior surface enhances energy efficiency and hygrothermal comfort of building envelopes while maintaining moisture equilibrium and preventing mold over long-term use [30]. Integrated envelopes maintain moisture balance by limiting conditions for condensation and fungal growth, such that relative humidity levels in critical locations do not exceed 75% during a three-year monitoring period [30]. An overview of impactful findings and statistics related to enhancing building envelopes with this novel strategy is outlined in Table 10.1.

Textile-Reinforced Mortar (TRM) coupled with thermal insulation represents a novel retrofitting solution aimed at the improvement of the seismic performance and energy efficiency of masonry buildings. Utilizing a mortar matrix and high-strength textiles for increased tensile strength and ductility in TRM or textile-reinforced mortar combined with reduced heat transfer thermally allows these buildings to better withstand seismic events. Rainproofing as a strategy integrates structural components with climate-specific solutions in the layout of the building envelope, serving both function and aesthetics [31, 32].

Vacuum Insulation Panels (VIPs) are a relatively new thermal insulation material, which provides significantly better insulation than conventional insulation products. There is growing interest regarding the use of VIPs in the renovation of old buildings, reflected by an exponential rise in related research publications since 2010.

Life cycle analyses demonstrate that fumed silica VIPs have a higher environmental impact, and costs compared to conventional thermal insulation materials based on thermal transmittance but still result in reasonable payback period. VIPs can be economically feasible over other types of insulation when considered for application on outer walls, (particularly when accounting for the added square footage area) placing VIPs as an effective insulation material for retrofitted old buildings [36]. The values of mean diameter and internal pressure of micro materials used for VIP are important parameters and in direct correlation with its thermal conductivity. The thermal conductivity can be reduced by reducing the mean diameters and internal pressures of micro materials, but if they are too low, they can increase the costs [37]. Therefore, further research is needed to optimize these properties for cost-effective and efficient VIP applications.

Wood Fiber Insulation (WFI) is a promising natural alternative that has been commercially successful in Europe since the end of the 1900s but is still ascending in the attic of its global scenarios. Material development research on WFI has mainly focused on cold climates and lacks comprehensive investigation into its biological and hygroscopic properties. However, existing research indicates that WFI performs effectively in wall and roof assemblies, offering benefits as a biologically sourced, vapor-permeable material that promotes better moisture regulation and healthier indoor air quality [38, 39]. Walls with wood fiber insulation components can present high moisture level risks to the wall as mineral wool insulation, however, wood fiber

Table 10.1 Overview of building envelope retrofitting upgrades

Aspect	Key findings	Challenges	References
Energy consumption in buildings	• Buildings contribute significantly to energy use and emissions • Poor insulation and outdated methods • Emerging materials underused in hot climates	• Proper application of advanced materials remains limited	[24–26]
Innovative envelope materials	• PCM, bio-based and hemp concrete improve comfort and energy • PCM near interior enhances performance	• Limited research on combined material use and long-term hygrothermal effects	[30]
Insulation and air sealing	• Older buildings have poor insulation, high air leakage • Insulation performance degrades over time	• Accurate measurement of insulation degradation often lacking	[27–29]
Seismic and thermal retrofit	• TRM + insulation improves seismic and energy performance—rainproofing integrates structural/climate	• Integration of seismic and thermal retrofit techniques is complex	[31–35]
Vacuum insulation panels	• VIPs provide superior insulation • Thermal conductivity depends on microporous properties	• High cost and environmental impact • Need for optimizing material properties	[36, 37]
Wood fiber insulation (WFI)	• Biologically sourced, vapor-permeable • Enhances moisture regulation and air quality	• Mainly researched in cold climates • Sensitivity to moisture	[38–40]
Challenges and future research	• Bio-based materials offer benefits like moisture buffering • Circular economy potential	• Moisture sensitivity—environmental trade-offs—limited social impact data	[41, 42]

insulation also has the benefit of storing condensation and dispersing it across a larger volume [40]. Installing an exterior air barrier with very high thermal resistance almost always reduces the wall construction moisture content and therefore decreases the risk of mold growth [40].

Although bio-based materials have many attractive benefits, including moisture buffering and very high vapor permeability, one of their main hurdles is their sensibility to moisture, which will rule the field for future work [41]. Circular bio-based building materials, which can be produced from renewable biological resources or, more broadly, from by-products originating from biomass, represent an interesting option to alleviate the negative impact of the building sector on the environment through the lens of a circular economy. While the most common assessments of the impacts of the HVAC occupations are environmental assessments (mostly through life cycle analysis), social impact assessments are less frequent. Moreover, some (or more) circular bio-based materials might face issues in other impact categories such as eutrophication and land use, calling into question their overall economic viability from a life cycle point of view [42].

10.3.2 Envelope Fenestration Upgrades

The performance of many historical buildings has deteriorated over time and does not comply with modern standards due to their age. Therefore, retrofitting has become a more economical and energy-efficient choice than demolition and reconstruction, especially for buildings with conservation value. Heat loss through the exterior windows of these buildings is one of the biggest challenges despite their historical appearance. The new window technologies with improved glazing can reduce heat loss and an improves thermal comfort [43], which can be effective solution in the energy efficient renovations. Through this method, they are enhancing energy efficiency whilst providing a model for all other renovations on external windows of historic buildings [44].

An empirical analysis of the thermal performance of *slim double-skin windows* of a historical building before and after the window remodel shows a more pleasant indoor heating experience during colder months and saves 34.2% of the energy consumed in buildings each year compared with conventional windows [45]. A retrofitting strategy on a representative 1960s single-family house in Germany, including new windows and integrated ventilation indicate an increase in indoor air temperature of 2.5 °C and a reduction in relative humidity by 3.8% due to medium-scale internal insulation retrofitting. For the home office, the results from the simulation showed heating energy consumption reduced by as much as 77%, a slight increase in CO_2 concentration, and higher indoor temperatures compared to pre-retrofitting. The new approach showed considerable potential to save energy and improve Indoor Environment Quality (IEQ), especially with respect to enhanced window operation and ventilation strategies [46, 47].

Electrochromic windows, also known as smart glass, are highly useful due to their energy economy and dynamic control of the visual and thermal parameters of space. These windows automatically adjust their tint in response to electrical signals. Commercialized electrochromic products have been studied extensively, with full-scale research identifying key parameters that define their optical, thermal, and electrical properties. Thanks to this technology, cooling energy demand can be reduced by approximately 25% and peak cooling loads by 13%. The window can provide privacy on demand and protect the interior from UV radiation. The costs per square meter are higher compared to the similar technologies while the installation requires professional expertise due to the electrical component involvement [48]. Despite limitations electrochromic windows are gaining popularity in sustainable architecture for their long-term energy savings and modern aesthetics [49].

Traditional smart windows make use of external sources of power to switch their color and switch the intensity of the light going to the inside, which only makes their use limited. Emerging technology is focused on self-powered electrochromic smart windows that can respond in real-time to changes in external light intensity, modulate the indoor light and heat levels without consuming additional electricity. Novel electrochromic materials were applied in hydrogel-state electrochromic devices with excellent performance, such as low voltage, fast response time, and high optical contrast [50].

A GA multi-objective optimization approach for optimizing the control parameters of smart shade curtains, as well as building performance standards using EnergyPlus software were performed regarding nine climates around the world. The results indicate that the inner curtain is clearly more effective than the outer curtain, with annual energy demand reductions of 4–30% and thermal and visual comfort improvement of 64–11% and 60–81%, respectively, based on the geographical aspect of the building [16].

10.3.3 HVAC Systems Upgrade

Older HVAC systems often operate under higher energy consumption and may provide poor air quality or inconsistent temperature regulation. Upgrading HVAC systems in retrofit buildings such as Variable Refrigerant Flow (VRF) systems, heat pumps, or energy recovery ventilators can significantly increase energy efficiency and reduce operational costs. This provides better control of indoor environments through zoning, smart thermostats, and integration with renewable energy sources.

Fan Coil Unit (FCU) systems have lower initial investment costs compared to Variable Refrigerant Flow (VRF) systems. VRF systems were 28% more efficient in terms of operating costs compared to FCU systems in a student dormitory setting [51, 52]. In addition, VRF systems provide a 24.6% advantage in investment costs, 32.5% in operating expenses, and 33.7% in energy consumption over FCU systems. Despite the higher initial costs, VRF systems are more cost-effective in the long run due to their energy efficiency and reduced operating expenses [53–55]. Implementation of evaporative cooling helps to increase in refrigeration system capacity in about 10–20% and reduce energy consumption about 3% [56].

The use of artificial intelligence (AI) technology for the optimal operation of HVAC systems is a most important feature that provides different types of algorithms, including supervised learning, reinforcement learning, and deep learning, in the fields of energy consumption prediction, protection of indoor environment, and fault detection. The synergy of AI with digital twin technologies by covering intelligent control, multi-objective optimization, and system diagnostics is investigated by the researchers [57]. Although promising HVAC optimization results have been reported, challenges remain to be addressed in practical applications, including model accuracy, generalization, and integration with existing systems [58, 59].

These findings are valuable to decision-makers to restore cooling systems in historical buildings. They offer a potentially replicable methodology that could benefit other cultural heritage organizations, designers, and engineers working on HVAC restoration projects. However, the uniqueness of the chosen case study gives it little universal significance. Moreover, using a primary decision-making matrix might neglect features relevant to different historical buildings and the preferences of possible stakeholders. Further studies should also explore other multiple benefit retrofitting pathways for a holistic assessment of the avoided emissions, including environmental footprints [60].

10.4 Lighting System Upgrade

Artificial lighting significantly contributes to electrical energy consumption in buildings, highlighting the need for effective lighting design strategies that reduce energy use while ensuring visual comfort. As historic and architecturally significant buildings increasingly become targets for energy efficiency improvements, the energy audit process must be tailored with a detailed understanding of their energy consumption profiles and include cost–benefit analyses to identify effective energy-saving opportunities without compromising heritage value. Different energy savings are achieved through the renovation of the lighting system in historical buildings, taking into consideration different luminaire types, arrangements, mounting heights, and wall reflectance values. In a case study from public offices in Italy, seven interventions on indoor and outdoor lighting were done in compliance with the value and show short payback periods between 4 and 34 months [61]. Different strategies of implementing the lighting system retrofit were analyzed to study their effects. Replacing luminaires is the only way to significantly reduce energy consumption, according to the research [62]. Using the lighting design software DIA Lux evo 7.0, the study simulates a real-scale room under different scenarios. Various design factors, such as surface finishing reflectance, luminaire type, quantity, and mounting height, were analyzed for their impact on lighting energy consumption and visual comfort. The results indicate that the type of luminaire greatly influences both the quantity and quality of indoor light, while enhancing surface reflectance can lead to energy savings of up to 45%, emphasizing the importance of integrating multiple strategies tailored to specific tasks and occupant activities [63]. Lighting is important aspect of occupant comfort. Appropriate lighting system can enhance the comfort and reduce fatigue by minimizing visual strain through established standards [64].

10.5 Water Conservation and Retrofitting Efforts

Water conservation and efficiency during retrofitting can lead to significant reductions in water usage. Implementing water-efficient techniques such as installing low-flow plumbing fixtures, showerheads, and toilets, can decrease water consumption by up to 50% without compromising performance.

Rainwater harvesting systems allows the use of rainwater for various non-potable applications such as irrigation and toilet flushing, which reduces reliance on community water supplies. Another effective strategy for water conservation is the reuse wastewater from sinks, showers, and washing machines for various non-potable applications. These systems are beneficial in instanced where water conservation is a priority. Implementation of water-saving technologies in retrofitting projects enhance sustainability and contribute to broader environmental conservation efforts [65, 66]. Some evidence suggests that retrofits of public buildings can achieve substantial water savings. To accelerate retrofit adoption and address core challenges, clear implementation frameworks are needed. A mixed-methods approach that integrates scenario modeling and stakeholders can effectively guide the development of water-focused retrofit strategies [67].

Old residential buildings equipped with rooftop photovoltaic also have a deep impact on water management as a considerable amount of energy is required to heat and distribute water. The energy consumed in these water-related processes could be reduced by utilizing the rooftop PV systems [68]. The integration of rooftop PV with solar water heating systems can help to optimize the use of energy by utilizing renewable energy sources for hot water requirements. Therefore, integrating rooftop PV with efficient water systems has the potential to yield energy and water benefits, making the system more resilient and resource-efficient [69].

Advanced Energy Management System (AEMS) that integrates rooftop solar photovoltaic (PV) systems and energy-efficient appliances to achieve energy optimization within households while also enabling users to sell excess electricity to the grid. The AEMS realizes substantial results of 28.59–35.48% cost savings over traditional systems, which are useful for residential energy management in Bangladesh. The installation of such innovative solutions while retaining modern building architecture will lead to more energy-efficient buildings and help to transition towards a sustainable energy future, both economically and environmentally [70].

10.6 Conclusions

The results highlight the importance of retrofitting to convert existing buildings into NZEB. The topics offered by this case study provide some solutions on how to upgrade the system to reach the desired point. Retrofitting aligns with the short and long-term objectives of sustainability by incorporating requisite design principles, new materials, and modern technologies in the retrofitting process maximizes not only occupant comfort but, more importantly, sustainability. The discussed case studies demonstrate that considerable energy and economic savings are possible, underscoring the need to integrate renewable energy solutions and to optimize building performance. The construction industry is under significant pressure to comply with strict environmental regulations, making this study a valuable reference for architects, engineers, and policymakers seeking to achieve NZEB standards for existing buildings. But with the construction sector under intensifying scrutiny to adhere to environmental benchmarks, the findings presented in this research offer a critical reference for architects, engineers, and policymakers alike, striving toward sustainable, energy-efficient built environments that tread lightly on natural ecosystems. It is crucial that future efforts focus on continued research and collaboration to optimize the potential of retrofitting practices to engage existing buildings in the transition towards a more sustainable energy culture.

References

1. D'Agostino D, Tzeiranaki ST, Zangheri P, Bertoldi P (2021) Assessing nearly zero energy buildings (NZEBs) development in Europe. Energ Strat Rev 36:100680. https://doi.org/10.1016/J.ESR.2021.100680
2. Pacheco R, Ordóñez J, Martínez G (2012) Energy efficient design of building: a review. Renew Sustain Energy Rev 16(6):3559–3573. https://doi.org/10.1016/J.RSER.2012.03.045
3. De Luca G, Ballarini I, Lorenzati A, Corrado V (2020) Renovation of a social house into a NZEB: use of renewable energy sources and economic implications. Renew Energy 159:356–370. https://doi.org/10.1016/J.RENENE.2020.05.170
4. Grillone B, Danov S, Sumper A, Cipriano J, Mor G (2020) A review of deterministic and data-driven methods to quantify energy efficiency savings and to predict retrofitting scenarios in buildings. Renew Sustain Energy Rev 131:110027. https://doi.org/10.1016/J.RSER.2020.110027
5. Ibrahim M, Harkouss F, Biwole P, Fardoun F, Ouldboukhitine S (2024) Building retrofitting towards net zero energy: a review. Energy Build 322:114707. https://doi.org/10.1016/J.ENBUILD.2024.114707
6. Harkouss F, Fardoun F, Biwole PH (2018) Multi-objective optimization methodology for net zero energy buildings. J Build Eng 16:57–71. https://doi.org/10.1016/J.JOBE.2017.12.003
7. El-Darwish I, Gomaa M (2017) Retrofitting strategy for building envelopes to achieve energy efficiency. Alex Eng J 56(4):579–589. https://doi.org/10.1016/J.AEJ.2017.05.011
8. Wang Z, Shen H, Deng G, Liu X, Wang D (2024) Measured performance of energy efficiency measures for zero-energy retrofitting in residential buildings. J Build Eng 91:109545. https://doi.org/10.1016/J.JOBE.2024.109545
9. Schueftan A, Aravena C, Reyes R (2021) Financing energy efficiency retrofits in Chilean households: the role of financial instruments, savings and uncertainty in energy transition. Resour Energy Econ 66:101265. https://doi.org/10.1016/J.RESENEECO.2021.101265
10. Park H, Park J, Kim S, Chang SJ (2024) Energy retrofit technology for modern and contemporary educational historical buildings considering preservation and utilization aspects. Energy Rep 11:3995–4007. https://doi.org/10.1016/j.egyr.2024.03.057
11. An YS et al (2023) Retrofit of renewable energy systems in existing community for positive energy community. Energy Rep 9:3733–3744. https://doi.org/10.1016/j.egyr.2023.02.055
12. Feng S et al (2024) Sustainable regeneration of waste polystyrene foam as cooling coating: Building cooling energy saving, CO_2 emission reduction and cost-benefit prospective. J Clean Prod 434:140361. https://doi.org/10.1016/J.JCLEPRO.2023.140361

13. Pertile V, De Stefani L, Scotta R (2018) Development and characterization of a system for the seismic and energy retrofit of existing buildings. Procedia Struct Integrity 11:347–354. https://doi.org/10.1016/j.prostr.2018.11.045
14. Yin P, Luo Z, Pouramini S (2024) Case study on multi-objective modified supply-demand-based optimization algorithm for energy-efficient building retrofitting. Sustain Cities Soc 114:105734. https://doi.org/10.1016/J.SCS.2024.105734
15. Alexakis K, Benekis V, Kokkinakos P, Askounis D (2025) Genetic algorithm-based multi-objective optimisation for energy-efficient building retrofitting: a systematic review. Energy Build 328:115216. https://doi.org/10.1016/J.ENBUILD.2024.115216
16. Liu P et al (2024) Multi-objective optimization of office building envelopes properties and Venetian blinds using NSGA-II to save energy consumption and enhance thermal and visual comfort. Case Stud Thermal Eng 64:105484. https://doi.org/10.1016/J.CSITE.2024.105484
17. Šuman N, Marinič M, Kuhta M (2020) A methodological framework for sustainable office building renovation using green building rating systems and cost-benefit analysis. Sustainability (Switzerland) 12(15). https://doi.org/10.3390/SU12156156
18. Sharbaf SA, Schneider-Marin P (2025) Cost-benefit analysis of sustainable upgrades in existing buildings: a critical review. Energy Build 328:115142. https://doi.org/10.1016/J.ENBUILD.2024.115142
19. Tetteh MO, Boateng EB, Darko A, Chan APC (2023) What are the general public's needs, concerns and views about energy efficiency retrofitting of existing building stock? A sentiment analysis of social media data. Energy Build 301:113721. https://doi.org/10.1016/J.ENBUILD.2023.113721
20. Mininni GM et al (2024) Increasing wellbeing through energy demand reduction for net zero: Citizen perceptions of co-benefits of local measures. Energy Res Soc Sci 118:103799. https://doi.org/10.1016/J.ERSS.2024.103799
21. Clemett N, Carofilis Gallo WW, O'Reilly GJ, Gabbianelli G, Monteiro R (2022) Optimal seismic retrofitting of existing buildings considering environmental impact. Eng Struct 250. https://doi.org/10.1016/j.engstruct.2021.113391
22. Noureldin M, Ali A, Memon S, Kim J (2022) Fragility-based framework for optimal damper placement in low-rise moment-frame buildings using machine learning and genetic algorithm. J Build Eng 54:104641. https://doi.org/10.1016/J.JOBE.2022.104641
23. Falcone R, Carrabs F, Cerulli R, Lima C, Martinelli E (2019) Seismic retrofitting of existing RC buildings: a rational selection procedure based on Genetic Algorithms. Structures 22:310–326. https://doi.org/10.1016/J.ISTRUC.2019.08.006
24. Costa-Carrapiço I, Raslan R, González JN (2020) A systematic review of genetic algorithm-based multi-objective optimisation for building retrofitting strategies towards energy efficiency. Energy Build 210:109690. https://doi.org/10.1016/J.ENBUILD.2019.109690
25. Penazzato L, Illampas R, Oliveira DV (2024) The challenge of integrating seismic and energy retrofitting of buildings: an opportunity for sustainable materials? Sustainability (Switzerland) 16(8). https://doi.org/10.3390/SU16083465
26. Ghazwani K, Beach T, Rezgui Y (2025) Energy retrofitting using advanced building envelope materials for sustainable housing: a review. Build Environ 267:112243. https://doi.org/10.1016/J.BUILDENV.2024.112243
27. Antonopoulos CA et al (2019) Choose an item. In: Wall upgrades for residential deep energy retrofits: a literature review. [Online]. Available: https://www.ntis.gov/about
28. Yun WS, Ryu W, Lee D, Seo H (2024) Energy-saving potential estimation of retrofitting aged buildings considering external wall insulation degradation. J Build Eng 94:110022. https://doi.org/10.1016/J.JOBE.2024.110022
29. Casamassima L, Kranzl L (2024) Exploring insulation levels to enable low-temperature heating in selected buildings in two European climate zones. Energy Build 316:114324. https://doi.org/10.1016/J.ENBUILD.2024.114324
30. Wu D, Rahim M, El Ganaoui M, Bennacer R, Liu B (2022) Multilayer assembly of phase change material and bio-based concrete: a passive envelope to improve the energy and hygrothermal performance of buildings. Energy Convers Manag 257:115454. https://doi.org/10.1016/J.ENCONMAN.2022.115454

31. Pohoryles DA, Bournas DA, Da Porto F, Caprino A, Santarsiero G, Triantafillou T (2022) Integrated seismic and energy retrofitting of existing buildings: a state-of-the-art review. J Build Eng 61:105274. https://doi.org/10.1016/J.JOBE.2022.105274
32. Yang P, Krevaikas T (2024) Experimental and analytical investigation of the tensile mechanism of textile-reinforced mortar (TRM) composites and fabric grids in elevated temperatures. Constr Build Mater 431:136495. https://doi.org/10.1016/J.CONBUILDMAT.2024.136495
33. Dehghan F, Porras Amores C (2025) Simulation-based multi-objective optimization for building retrofits in Iran: addressing energy consumption, emissions, comfort, and indoor air quality considering climate change. Sustainability 17(5):2056. https://doi.org/10.3390/SU17052056
34. Pohoryles DA, Bournas DA (2020) Seismic retrofit of infilled RC frames with textile reinforced mortars: state-of-the-art review and analytical modelling. Compos B Eng 183. https://doi.org/10.1016/j.compositesb.2019.107702
35. Truong GT, Kim JC, Choi KK (2017) Seismic performance of reinforced concrete columns retrofitted by various methods. Eng Struct 134:217–235. https://doi.org/10.1016/j.engstruct.2016.12.046
36. Božiček D, Peterková J, Zach J, Košir M (2024) Vacuum insulation panels: an overview of research literature with an emphasis on environmental and economic studies for building applications. Renew Sustain Energy Rev 189:113849. https://doi.org/10.1016/J.RSER.2023.113849
37. Mao S, Kan A, Huang Z, Zhu W (2020) Prediction of thermal performance of vacuum insulation panels (VIPs) with micro-fiber core materials. Mater Today Commun 22:100786. https://doi.org/10.1016/J.MTCOMM.2019.100786
38. Künzel HM (2022) Characteristics of bio-based insulation materials. Green Energy Technol 163–175. https://doi.org/10.1007/978-3-030-98693-3_6
39. O'Brien L, Li L, Friess W, Snow J, Herzog B, O'Neill S (2025) Review of bio-based wood fiber insulation for building envelopes: Characteristics and performance assessment. Energy Build 328:115114. https://doi.org/10.1016/J.ENBUILD.2024.115114
40. Bunkholt NS, Rüther P, Gullbrekken L, Geving S (2021) Effect of forced convection on the hygrothermal performance of a wood frame wall with wood fibre insulation. Build Environ 195:107748. https://doi.org/10.1016/J.BUILDENV.2021.107748
41. Weerasinghe LNK, Darko A, Chan APC, Blay KB, Edwards DJ (2024) Measures, benefits, and challenges to retrofitting existing buildings to net zero carbon: a comprehensive review. J Build Eng 94:109998. https://doi.org/10.1016/J.JOBE.2024.109998
42. Chen L et al (2024) Biomaterials technology and policies in the building sector: a review. Environ Chem Lett 22(2):715–750. https://doi.org/10.1007/S10311-023-01689-W
43. Ascanio-Villabona J, Orejarena-Osorio N, Castillo-Leon NY, Tarazona-Romero BE, Jaimes-Quintero KT (2024) Thermal comfort evaluation using different glazing technologies in an energy model of rural housing in a cold climate. Herit Sustain Dev 6(2):735–748. https://doi.org/10.37868/HSD.V6I2.922
44. Qin X, Liu H, Zhang X, Jiang N, Yang L, Jin X (2024) Thermal analysis of the window-wall interface for renovation of historical buildings. Energy Build 310:114108. https://doi.org/10.1016/J.ENBUILD.2024.114108
45. An Y, Choi H, Kim E, Kang K, Kim S, Kim T (2022) Empirical analysis of the effects of window remodeling using the slim double-skin window on the thermal performance of old buildings. Energy Build 277:112581. https://doi.org/10.1016/J.ENBUILD.2022.112581
46. Ortiz M, Itard L, Bluyssen PM (2020) Indoor environmental quality related risk factors with energy-efficient retrofitting of housing: a literature review. Energy Build 221:110102. https://doi.org/10.1016/J.ENBUILD.2020.110102
47. Pungercar V, Zhan Q, Xiao Y, Musso F, Dinkel A, Pflug T (2021) A new retrofitting strategy for the improvement of indoor environment quality and energy efficiency in residential buildings in temperate climate using prefabricated elements. Energy Build 241:110951. https://doi.org/10.1016/J.ENBUILD.2021.110951

48. Doors and windows with smart glass technology: costs and benefits explained. Accessed 30 May 2025. [Online]. Available: https://www.panda-windows.com/blog/doors-and-windows-with-smart-glass-technology-or-cost-pros-and-cons?utm_source=chatgpt.com
49. Sibilio S et al (2016) A review of electrochromic windows for residential applications. Int J Heat Technol 34(Special Issue 2):S481–S488. https://doi.org/10.18280/IJHT.34S241
50. Li YZ, hao Cui Z, yun Jiang D, yang Tang P, bin Gong C, Tang Q (2025) Dynamic optical and thermal modulation of electrochromic smart windows and sunglasses based on benzothiadiazole-extended viologen derivatives. Dyes and Pigments 233:112543. https://doi.org/10.1016/J.DYEPIG.2024.112543
51. Bai L, Tan Z (2024) Optimizing energy efficiency, thermal comfort, and indoor air quality in HVAC systems using a robust DRL algorithm. J Build Eng 98:111493. https://doi.org/10.1016/J.JOBE.2024.111493
52. Cui W, Liu G, Wang Y, Li K (2024) Integrated optimization of the building envelope and the HVAC system in office building retrofitting. Case Stud Therm Eng 62:105185. https://doi.org/10.1016/J.CSITE.2024.105185
53. Variable refrigerant flow (VRF) systems: an emerging technology. Accessed 30 May 2025. [Online]. Available: https://studylib.net/doc/18704895/variable-refrigerant-flow--an-emerging-air-conditioner-an...?utm_source=chatgpt.com
54. Adigüzel N, Verimliliği E, Etkisinin Ü, Analizi K (2024) Comparative analysis of the effect of replacing central system heating and cooling with Vrf technology on energy efficiency. J Energy Trends 1(1):10–20. https://doi.org/10.5281/ZENODO.12566525
55. Özahi E, Abuşoğlu A, Kutlar Aİ, Dağcı O (2017) A comparative thermodynamic and economic analysis and assessment of a conventional HVAC and a VRF system in a social and cultural center building. Energy Build 140:196–209. https://doi.org/10.1016/J.ENBUILD.2017.02.008
56. Hussain HM, Hammdi SH (2024) Numerical simulation of integrating an air conditioner with an evaporative air cooler. Herit Sustain Dev 6(1):43–66. https://doi.org/10.37868/HSD.V6I1.346
57. Heti K, Silvanskaya G, Khambir V, Harkava V, Ivanchenko K (2025) The analysis of engineering adaptation to global challenges and strategic planning for the future. Sustain Eng Innov 7(1):137–146. https://doi.org/10.37868/SEI.V7I1.ID453
58. Lu S et al (2025) Exploring the comprehensive integration of artificial intelligence in optimizing HVAC system operations: a review and future outlook. Results Eng 25:103765. https://doi.org/10.1016/J.RINENG.2024.103765
59. Jain R et al (2023) Optimization of energy consumption in smart homes using firefly algorithm and deep neural networks. Sustain Eng Innov 5(2):161–176. https://doi.org/10.37868/SEI.V5I2.ID210
60. Etemad A, Zare N, Shafaat A, Bahman AM (2024) Assessing strategies for retrofitting cooling systems in historical buildings. Energy Rep 11:1503–1516. https://doi.org/10.1016/J.EGYR.2024.01.017
61. Salvadori G, Fantozzi F, Rocca M, Leccese F (2016) The energy audit activity focused on the lighting systems in historical buildings. Energies (Basel) 9(12). https://doi.org/10.3390/EN9120998
62. Ciampi G, Rosato A, Scorpio M, Sibilio S (2015) Retrofitting solutions for energy saving in a historical building lighting system. Energy Procedia 78:2669–2674. https://doi.org/10.1016/J.EGYPRO.2015.11.343
63. Makaremi N, Schiavoni S, Pisello AL, Cotana F (2019) Effects of surface reflectance and lighting design strategies on energy consumption and visual comfort. Indoor Built Environ 28(4):552–563. https://doi.org/10.1177/1420326X18793170
64. Alketbi MAS, Mushtaha ES (2025) The impact of artificial lighting on improving the visual experience of museum visitors: a case study of Sharjah Museum of Islamic Civilization. Herit Sustain Dev 7(2):919–940. https://doi.org/10.37868/HSD.V7I2.1271
65. Retrofitting historic buildings with modern water management systems—Nexteel Industries Private Limited. Accessed 30 May 2025. [Online]. Available: https://nexteel.in/retrofitting-historic-buildings-with-modern-water-management-systems/?utm_source=chatgpt.com

66. Ahmed W, Alazazmeh A, Asif M (2023) Energy and water saving potential in commercial buildings: a retrofit case study. Sustainability 15(1):518. https://doi.org/10.3390/SU15010518
67. Bertone E et al (2018) Guidelines, barriers and strategies for energy and water retrofits of public buildings. J Clean Prod 174:1064–1078. https://doi.org/10.1016/J.JCLEPRO.2017.11.065
68. Borodiņecs A, Lebedeva K, Odineca T (2024) Evaluation of household electricity consumption in multi-apartment buildings for optimization of rooftop PV systems. Energy Build 325:114971. https://doi.org/10.1016/J.ENBUILD.2024.114971
69. Wang P, Yu P, Huang L, Zhang Y (2022) An integrated technical, economic, and environmental framework for evaluating the rooftop photovoltaic potential of old residential buildings. J Environ Manage 317. https://doi.org/10.1016/j.jenvman.2022.115296
70. Shufian A, Fattah SA (2024) Integrated rooftop solar PV-based residential advanced energy management system: an economic involvement of energy systems for prosumers. Cleaner Energy Syst 9:100150. https://doi.org/10.1016/J.CLES.2024.100150

Chapter 11
Tools and Resources for ZEBs Development

Abstract The architecture and construction sectors have been driven towards innovative design aimed at reducing energy demand in whic zero energy buildings became a leading paradigm for sustainable solutions. This chapter explores tools for performance forecasting and process optimization from the early stages of building design. Particularly, tools including energy performance assessment models, machine learning techniques, and dynamic simulation engines to direct energy efficiency optimization from conceptual to operational phases. The impact of educational programs, accreditation systems and the industry co-operation on spreading expertise and acceptance of zero energy buildings principles were also discussed. Presented case studies examine some of the best practices, challenges and opportunities that accompany zero energy buildings implementation.

Keywords Building energy performance · ZEB design tools · Training programs · Educational programs · Sustainability

Abbreviations

AI	Artificial intelligence
ANN	Artificial neural network
BEM	Building energy modeling
BIM	Building information modeling
BME	Biomass energy
BMS	Building management systems
BUS	Build up skills
CDA	Conditional demand analysis
COTE	Committee on the environment
GTE	Geothermal energy
HVAC	Heating, ventilation, and air conditioning
IoT	Internet of things
LED	Light emitting diodes

B. Duraković, *Zero Energy Buildings*, Green Energy and Technology,
https://doi.org/10.1007/978-3-032-16027-0_11

NN	Neural network
PV	Photovoltaic
RES	Renewable energy sources
VET	Vocational education and training
ZEB	Zero energy building

11.1 Introduction

Growing concerns about climate change and energy conservation have encouraged the architectural and construction sectors to pursue innovative approaches to improving buildings while minimizing energy consumption. This evolving focus has prompted extensive research into groundbreaking strategies, with the Zero Energy Buildings (ZEBs) concept emerging as a leading paradigm. These are high performance buildings with a goal of using as little energy as possible in operation, while matching the energy loads with energy generated either on or off site in the form of renewable energy. ZEBs offer substantial long-term economic and environmental benefits such as low operating costs, improved energy security, and low carbon emissions [1].

Even though ZEBs provide substantial environmental advantages, the design and construction of these advanced structures remain complex and challenging. One common challenge is the fact that it is more expensive to design and build, which may add up to 15% on to the costs of construction [2]. This additional investment is mainly because of the requirement of enhanced structure materials and systems, energy efficiency, and the integration of renewable energy sources [3].

To facilitate the development and practical application of ZEBs there are tools and methods for the assessment of the building energy performance. These tools perform energy simulations combined with machine learning techniques to forecast system performance and enhance energy efficiency from the initial phases of design [4].

The purpose of this chapter is to discuss various aspects of ZEB design and development from energy simulation techniques to renewable energy sources. It also examines commercially available tools used for ZEB design and analysis, examines the role of educational and training programs in developing competent human capital, and examines the contribution of industry associations in promoting collaboration and innovation [5]. While presenting the various successful case studies, tools and practices, this chapter provides a general insight into the ZEB concept and ho w it can significantly transform the global building environment.

11.2 Mathematical Models for ZEB Design and Analysis

ZEB design is focused on achieving optimal energy efficiency within buildings, aiming to minimize energy demand and the operational costs associated with energy use [6]. The concept of reaching ZEB standards entails additional costs in construction that may also rise to 15% [2]. However, an excellent energy performance level which is comparable to the level of ZEBs may be achieved at reasonable cost through appropriate design decisions searching for the best compromise between utilizing energy and generating it. For instance, somewhat higher energy use might be tolerable to offset the cost of construction due to the building envelope, heating, ventilation and air conditioning (HVAC) equipment.

When designing and conceptualizing ZEBs, energy simulation is an essential tool that helps planners, engineers, and architects maximize energy efficiency prior to construction [7]. Designers can make better-informed design decisions by identifying the building components that have the biggest effects on energy efficiency through the simulation of energy usage and production. Additionally, these models aid in the decision-making process for both passive and active design techniques, including window placement, insulation, and shading device integration. Stakeholders can guarantee that the building satisfies its energy efficiency objectives and saves expensive post-construction alterations by modeling energy performance beforehand [8].

Methodologies for energy modeling are frequently divided into top-down and bottom-up approaches, each of which has unique benefits. Stakeholders may strike a compromise between precise building element optimization and broad energy goals thanks to this dual approach. Buildings that are more energy-resilient and self-sufficient can result from their ability to integrate renewable systems more intelligently [9]. When combined, these approaches guarantee that ZEBs are economical, ecologically friendly, and energy efficient right from the start.

11.2.1 Top-Down Approaches

In the field of energy consumption analysis, top-down approach provides flow of vision briefly and look at large population of the buildings and their relationship with the economic, social and physical environment factors. These approaches are particularly useful for predicting energy requirement on a larger scale, city, regional or even the national level. They help to explain the ways in which alterations in the macroeconomic environment, often consisting of technology and demography, might affect energy levels. For urban planners, policymakers and other stakeholders, these models provide key information to better anticipate and address energy demand.

Top-down models are based on statistical inference and are classified based on the elements to be quantified. This approach examines the total stock of a group of buildings and the interrelationships between energy and a wide-ranging number of factors such as the economy, population, technology or climate [10]. In other words, top-down strategies describe the dependency between energy utilization and its principal determinants.

Therefore, top-down models include only macro data, and this is their advantage as well as disadvantage. These models can contain historical data and generate extremely accurate models displaying accumulated tendencies for future markets, reflecting the macroeconomic influences on energy consumption. For example, the econometrical model to evaluate housing consumption of energy over time [11]. But they fail to give good forecasts when there is a large shift in the economy, the technology or the climate. Other top-down approaches for energy consumption estimation have been derived for several countries including the USA [12, 13], Norway [14], Denmark [15], Türkiye [16] and New Zealand [17].

11.2.2 Bottom-Up Approaches

Bottom-up approaches focus on various attributes of building characteristics, systems, materials and occupancy patterns to give accurate energy demands. From an engineering perspective, it is easier to incorporate appliances into buildings and even retrofit existing structures using current methods with decreased energy demand and thus more environmentally friendly. These models are associated with finer grains of building envelope differentiation such as system and installation, geometry form, thermal performance characteristic of building construction material, external conditions, internal set point temperature, occupants, and hours [18]. Because the knowledge level of a building is very detailed, due to the accuracy of the estimate, it becomes possible to come up with a quite reliable evaluation of the energy demand at the level of single building.

Bottom-up approaches are also shown to be distinguishable into two substantially different categories such as quantitative and quantitative techniques [19]. The required constraints are estimated through statistical inference since it establishes correlation between construction descriptors and their energy consumption. These approaches operate as do the previously mentioned top-down strategies, which have the benefit of applicability and simplicity despite the use of disaggregated data. Bottom-up models include e.g. regression, conditional demand analyses and artificial neural network. Regression analysis is a safe and time-efficient method for the evaluation of energy consumption in buildings. It is premised on the estimates of the regression coefficients considered to have an impact on the energy use of a building. A detailed understanding on how various regression analyses can be used to forecast energy consumption in the residential sector is provided in the work by Fumo and Rafe Biswas where the authors make comparison between simple, multiple and multivariate linear regression with the non-linear regression models [20].

Engineering-based methods estimate, with the support of simulation engines, the energy consumption of the building through the examination of the thermophysical properties that define how the building as the thermodynamic system behaves with the outdoor environment [21]. Engineering methods offer the most precise approximations, but they necessitate direct, in-depth knowledge of the specific buildings under evaluation. Accurate assessment depends on data about the building geometry, materials, systems, and occupancy behavior. In more detail, engineering methods are divided into steady state building energy assessment tools, dynamic building energy simulation and other building energy assessment tools. Steady state building energy assessment tools are mainly applied in the energy certification processes of buildings. These tools are generally developed on a physical approach allowing users to minimize input data and reduce computation time.

Dynamic building energy simulation tools primarily would like to do a precise estimation of the thermodynamic characteristics of a building structure inclusive of the simulating of HVAC systems. These tools could be highly accurate, but to achieve this they demand much time and a rigorous filtration of input data. Besides that, they also need long simulation sessions, and powerful computer processors to execute tasks. These types of energy assessment tools fall under the building energy modeling which is core in building energy design and assessment task related to external and internal gains. It is also necessary to possess a concrete, scientific, knowledge of the climatic area of the building locations, including microclimate and local impacts [22].

11.2.2.1 Conditional Demand Analysis

Conditional demand analysis (CDA) is a unique statistical method which proves very effective in explaining and analyzing energy demand in buildings which a definition of special interest, zero energy buildings (ZEBs). This method uses regression analysis to disaggregate energy use in a building showing, in detail, how different end use appliances are contributing to energy use. The energy requirement is stated in a sectorized additive way that considers specific operations and behavior of each device in conjunction with occupancy activity, equipment traits, and local climate. CDA approaches are especially appropriate for detecting variability in energy consumption patterns that depend on temperature settings, occupancy levels and operational profiles of devices, etc. For example, the use of more specialized modeling such as that used in the analysis of nearly zero energy buildings allows for an understanding of how specific aspects of design such as levels of insulation and characteristics of windows continue to have an influence on levels of energy demand [23]. This underlines the significance of accuracy in quantitative data collection and subsequent analysis of the factors that underpin energy consumption profiles. While common in CDA methodology, data collection is typically through surveys and records from utility companies that contain information about the operational profile of the devices and power consumption rates connected thereto. Such an approach allows for the definition of certain factors that affect energy consumption, and thus, the energy consumption can be forecasted with higher precision.

11.2.2.2 Artificial Neural Network

Artificial Neural Networks (ANNs) represent as a major step forward in the simulation of building energy which is equally useful in ZEBs. What these mathematical models provide is the ability to have detailed learning and adaptation processes which are based on the connections of a biological brain. ANN techniques for modeling energy consumption started in the early 1990s and this is considered as the first-time computation intelligence is incorporated into energy analysis [24]. It has been widely used in energy simulation, especially for the ZEB design analysis, in consideration of balanced energy consumption and the local renewable energy generation for these buildings. By training ANN on historical data that incorporates data on occupancy, climate, and other products used, the models are capable of estimating energy demands. For instance, one work has demonstrated that ANNs can estimate the annual electrical energy consumption in nearly zero energy building designs by using data collected from building and scalar simulation methodologies [25]. This capability helps architects and engineers to enhance building designs based on energy efficiency and sustainability indicators. The assessment of energy efficient strategies for the ZEB design is where energy simulation methods have considerable value. Table 11.1 provide a summary of techniques required for energy demand prediction, along with their advantages and disadvantages.

Selecting the right energy simulation approach for ZEB design and assessment depends on many factors including the project needs and data availability. Large scale and detailed investigations are performed through the top-down and bottom-up methods while narrow and short term are done through engineering and statistical methods. New technologies like artificial neural networks are good when it comes to modeling demand with a big data set. All the above-mentioned have some benefits in enhancing energy efficiency and sustainable building.

11.3 Advancing Renewable Energy Integration

Renewable energy (RE) is now gradually emerging as the global model for efficient power generation for the future due to increasing commitments and sustainable development goals across nations. Out of all the renewable energy technologies, wind, solar, biomass and geothermal understandings have been identified as the key enabling technologies altering how energy can be generated, utilized and incorporated into the utility system. They contribute significantly to streamlining greenhouse emissions, increasing utility of energy, and bolstering energy security power. In addition, they create new opportunities for the economy because they stimulate the development of new ideas and the emergence of new types of income generating ventures. As in other developing countries, there therefore exists great potential for boosting sustainable energy projects where innovative technologies offer new opportunities [26–28].

Table 11.1 Overview of approaches and methods used in ZEB design and analysis

Method	Description	Strengths	Weaknesses
Top-down approaches	• Less covers broad system level aspects of buildings and energetic usage, large scale economic and social factors (demographic, policy, economy)	• It is very helpful for the purpose of forecasting energy demand at city or regional or national level • Develops the general understanding of the situation	May not deliver building level changes in detail; weak in anticipating technological or climate changes
Bottom-up approaches	• As a weakness: decisions may not be granular enough to reflect changes in the building level, because they do not foresee shifts in technology or climate	• Capable of quantifying energy requirements using information peculiar to each building; useful for single applications	• A lot of data must be sampled; very time-consuming for the computer
Statistical methods	• Predicts energy demand from building features using computational models such as the regression model	• Saves time, easy to design and implement, suitable for installation in nearly all construction structures	• The approach is not as specific as the engineering-based approach that is used in practice. May fail to incorporate some dynamics such as occupancy behavior
Engineering methods	• Because of the identified thermophysical parameters and interaction with the environment, it reproduces the energy consumption of the building	• Very precise; considers features of the construction and various systems	• Demands major sophisticating and preputial data input; the most time-consuming and power-sharply model
Conditional demand analysis	• An analytical technique that disaggregates energy usage into distinct types of loads, for example, cooling systems, lamps	• Orders energy consumption by individual devices, particularly useful for looking at opportunities for increasing efficiency	• Based on surveys and utility data; does not consider all factor influencing factors
Artificial neural networks	• Models the energy demand based on the principles of neural network and learn inputs from the data	• Has ability to work with many patterns and get better with time as more data is encountered	• Needs a lot of samples; it is often slower to perform and less easy to explain

New technologies provide opportunities to improve the feasibility of RE systems, whereas physical improvements help extend the adoption and use of RE systems. Integration of renewable electricity sources into the national and regional power systems requires comprehension of absolute, relativistic, conditional and total parameters influencing electricity use. Reduced-complexity top-down approaches offer a systematic framework for examining the socioeconomic, environmental, and technological drivers of energy consumption, enabling better exploitation of this potential.

These approaches of large-scale modeling and strategic planning are insufficient to deal with problems specific to a project. It is here that the use of bottom-up methodologies is appropriate, as these offer a granular view of how energy increases in defined areas or at the building level may be achieved. When used together, these methods form a comprehensive approach for handling energy issues on a small-scale helping design cities with improved energy efficiency. The main challenges and future perspectives for the successful integration of renewable energy systems are presented in Table 11.2.

Referring to Table 11.2, it can be observed that integration of renewable energy systems is a complex task involves many issues, technical, economic and social. It is important to target those policies that will promote the development of renewable energy resources as well as the development of infrastructure and increased public awareness that will help achieve sustainable urban developments.

11.4 Educational and Training Programs

With the advancement of ZEBs concept, there is a growing requirement of a qualified workforce capable of designing, constructing and managing these low energy buildings. There is great reliance in educational resources and training programs for professionals that can be effective in facing the complexities of ZEB design, enhanced energy efficiency and RE integration. These programs are relevant to a variety of audiences, including architects and engineers, construction project managers, policy makers and others within the built environment.

As the European Commission's (EC's) Build Up Skills (BUS) investigation suggests, there is a need for training the equivalent of 3 million workers and substantial improvement of Vocational Education and Training (VET) systems in several countries. The effectiveness of interventions towards higher energy efficiency in buildings relies on, amongst others, sufficiently skilled personnel [29]. The EU goals on increasing the energy efficiency of buildings suppose that workers who are involved in the construction industry will have enough knowledge to fulfil the requirements to ZEB.

Table 11.2 Challenges and future perspectives

Categories	Challenges	Future perspectives	References
Cost	Constructing and has been proven to increase the costs of developing ZEBs by up to 15%	It is also feasible that cost reduction can be achieved at the right time and proper strategy at the time of designing part	[2]
Education and human resources	Lack of skilled human resources in sufficient numbers	Recruiting more skilled professionals through academic programs, online courses and aptitude-oriented training	[29, 30]
Renewable energy utilization	Insufficient connections and expensive investment requirements for building ties between energy systems such as wind, solar, and biomass	Promoting the production and absorption of renewable energy resources by technological innovation and enhanced physical networks	[27, 31]
Political and economic factors	Insufficient adoption of the policy and financial incentives	Encouraging the use of ZEB through right policies and regulatory instruments, and proper support of local authorities	[32]
Technological development	Complexity and high learning curve of energy simulation and modeling tools	Accurate tools based on artificial intelligence and machine learning for the energy sector to make energy performance predictions within a short time	[20, 25]
Social acceptance	Lack of awareness and misunderstanding of the ZEB Concept	By increasing awareness through educational programs and presenting well-implemented pilot projects	[33, 34]

11.5 Online Courses and Certification Programs

Professionals are more in need of easily available and adaptable learning solutions since the energy industry is rapidly expanding [35]. As for the further development of skills, it is possible to turn to online courses as well as certification programs adopted for energy-saving construction systems and ZEB designing [36]. Target consumers of these programs are people who cannot attend standard formal learning programs at universities for various reasons. They also offer a way for those who are currently employed in engineering, construction, or energy industries to know the latest in their industries. Since it is possible to pursue education alongside full employment, few professions present barriers that limit the access to continued education at the same time as they maintain their professional positions [35]. Furthermore, these certification programs provide participants with certification enhancing one's employment opportunities. Online courses help to continue education, allowing an increasing demand for practical and professionally oriented education. Combining the interactive simulation, project-based evaluation with real-world case enhances learning more effectively and provides participants with real life design experience on ZEB and energy efficient building systems [36]. Finally, this method contributes to the industry changes in lean construction methodologies by graduating competent students to fit into the emerging need for energy efficient design and sustainable structures.

11.6 Industry-Specific Workshops and Training

Trade associations, manufacturers and other specialist bodies often offer such courses, which offer participants hands on experience and equip them with the latest trends and technologies in energy efficient building structures and renewable energy systems [30]. Such workshops contribute to narrowing the gap between theoretical knowledge and practical skills by focusing on technical issues of installation, maintenance, and application [37]. This approach is especially useful in making professionals immediately use new skills in projects they are conducting which makes them more efficient in the field [38].

Standard-based certification programs are an integral component of these training endeavors because participants receive certifications that increase their employability. For instance, the industry honors and acknowledge certain approvals like Leadership in Energy and Environmental Design (LEED), Passive House Institute US (PHIUS), and Building Performance Institute (BPI) [30], where LEED is the most widely used green building rating system. These programs allow professionals to prove their mastery of efficient design, passive construction techniques, and performance assessments of buildings; they can greatly help in positioning professionals as more favorable in the market. To ensure that the participants are informed of the current development, trends and practices only the technical experts in that field are used to facilitating the workshops [37]. By doing this trainee are better placed to meet the demands of the industry and contribute more to sustainable building projects. Table 11.3 outlines key educational pathways that support the development of skills and knowledge in ZEB design and sustainable architecture, tailored to various learner profiles and professional needs.

Referring to Table 11.3 it is observed that the diverse educational strategies available for advancing expertise in ZEB design and sustainable architecture are available. It shows how training programs, online certifications, and industry-specific workshops are tailored for different audiences including university students to work professionals and offering theoretical foundations and practical skills. Training programs emphasize a strong scientific background combined with hands-on projects, while online courses provide flexible access to up-to-date knowledge and certification opportunities for those unable to attend in person. Industry-specific workshops, such as LEED and PHIUS certifications online give the professionals

Table 11.3 Educational pathways for ZEB and sustainable design expertise

Educational method	Target audience	Content and advantages	Outcomes and certifications	References
Online courses and certifications	• For the specialists, who would like to update their knowledge in the respective lines of work • People who cannot attend physical education facilities	• Virtual hazards and assessments for energy systems and ZEB design and innovation • Existing actual day design exposure	• It LIS for LEED, PHIUS, BPI and other related certifications • It has improved career advancements and capacities	[35, 36]
Industry-specific workshops and training	• Individuals working in the area of construction, engineering and energy	• Expertise in technical information and practical experience reasonably related to the installation, maintenance or application of specialized equipment or processes • Target on novel technologies and phenomena affiliated to energy-efficient architecture	• You get a certification like LEED, PHIUS and BPI among others among the participants • Develop competencies in order to be useful in the process towards sustainable building projects	[30, 37]

convenient and beneficial opportunities to learn about them by focusing on real-world applications and emerging technologies in energy-efficient construction. Overall, these educational pathways play a crucial role in building a skilled workforce capable of contributing to sustainable building practices and meeting global energy efficiency goals.

11.7 Industry Groups and Collaboration Platforms

For construction professionals to be active practitioners in ZEB projects they are required to be members of a community of practice within the industry that embody digital platforms to share knowledge and resources. The most important role in this regard is served by trade associations as well as networking opportunities which are valuable sources of resources and contacts that can assist in boosting the effectiveness

Table 11.4 Key industry associations and networking for ZEB professionals

Organization	Role	Activities	Benefits to professionals
American Institute of Architects	Advocates sustainable design, supports energy-efficient and zero-energy buildings	Webinars, case studies, research publications, annual conferences, expos, local programs	Networking, knowledge sharing, learning about ZEB implementation, professional development
U.S. Green Building Council	Leading advocate for sustainable architecture, developer of LEED certification system	Conferences (e.g., Greenbuild), seminars, regional chapters, networking groups	Access to certification knowledge, peer interaction, exposure to ZEB innovations
Passive House Institute US	Promotes passive house standards as a pathway to zero-energy status in North America	Meetings, webinars, seminars, certification programs	Collaboration with architects, engineers, builders; knowledge exchange
International Passive House Association	Global promoter of Passive House standards	Discussion forums, international and regional conferences	Idea exchange, networking with specialists, insights into Passive House and ZEB
International Energy Agency	Focus on energy-efficient buildings and communities, including zero-energy buildings	Annual meetings, workshops, conferences	Updates on emerging trends, professional awareness, collaboration opportunities
Net Zero Energy Coalition	Peer network of stakeholders advancing zero-energy buildings	Workshops, conferences, best practice papers, learning materials	Access to latest technologies, policy discussions, community support, networking

of ZEB initiatives. Professional bodies offer forums whereby people in a certain field meet to exchange ideas, lessons, experiences, and developments of ZEB design and construction. For instance, looking at organizations such as the International Living Future Institute and the US Green Building Council, individuals can access certifications and guidelines that advocate for ZEB normally referred to as zero-energy building. Some of these associations hold annual conferences, workshops and seminars that present academic and industry professionals to discuss new information, findings and the practical implementation of ZEB technology [39, 40].

Substantial knowledge can be gained from these events not only on an individual level but also creates opportunities for partnerships that contribute to effective project implementation. Attractiveness to industry groups offers the most benefits for new members of the industry. Through the forums, young architects, engineers, and construction managers can learn from practitioners about the implementation of solutions based on ZEB projects' challenges and prospects. And through the training and capacity building programs that are given by these associations, new professionals' competence can be enhanced to meet the practical challenges of sustainable buildings [34, 41].

The *American Institute of Architectures* (AIA) has a considerable influence in the campaign for sustainable design through its Committee on the Environment (COTE). This committee actively supports the concept of energy-efficient and zero energy. AIA offers webinars, case studies, and research publications for its members. For instance, the architectural professional organization AIA is a key producer of information on sustainable design and ZEBs. The AIA annual Conference on Architecture and Design has breakout sessions that are related to sustainability and ZEBs and enables professionals to participate in related business and learning activities. There are a lot of networks to be made through AIA. The organization arranges expos that comprises sessions on ZEBs, sustainability and so forth, it creates a platform whereby specific personalities within the architectural field get to meet and learn while developing new concepts and ideas on projects. AIA local offices also host programs within which the members can meet with individuals from the sustainability as well as zero-energy building industries.

The U.S. Green Building Council (USGBC) is one of the most influential advocates for sustainable architecture and is responsible for developing the LEED certification system. In its advocacy positions, education, and certification programs USGBC supports the concept of zero-energy buildings [33]. The buildings which have attained LEED certification have been found to have high potential of efficiency in energy usage [42]. With the help of USGBC organizational connections, (including the Greenbuild International Conference & Expo), professionals can directly interact with industry peers, learn more about the principles of zero-energy buildings during specialized seminars, and potentially find partners [43]. It is a knowledge base on new technologies, strategies, and innovations related to the pursuit of increasing the number of zero-energy buildings [44]. However, this is also true since USGBC has given a nod to the creation of regional chapters and networking groups whereby interested individuals who are engaged in different but related projects and initiatives can be able to meet [45].

The *Passive House Institute US* or PHIUS remains at the forefront of advocating for passive house standards as an important way of achieving the zero-energy status in North America. The various steps used in the PHIUS certification help to make the construction of buildings efficient by using less power in heating and cooling [46]. PHIUS provides professional and social collaboration through meetings, webinars, and learning seminars that cover improved principles of energy-efficient construction. Those who become members of PHIUS or have the chance to attend events of this certification can connect to working professionals who are architects, engineers, builders, and others focused on creating structures that are energy-efficient or even zero-energy ones. These interactions enhance synergy and exchange of information, knowledge, and experience—critical to zero-energy design and deployment [47].

As a result, the great idea of creating zero energy buildings is a result of teamwork and industry knowledge, innovation and related professional people who are looking forward to having better and improved architectural designs. These groups are crucial not only in offering the information, support, and contacts necessary to thrive in, but also to simply keep pace in this growing and dynamic field. Through collaboration with such organizations, the professionals can participate in the shift of a global built environment to utilize built environment that requires zero energy and assist in the promotion of zero-energy buildings all over the world.

The *International Passive House Association* (IPHA) is responsible for the popularization of Passive House standards all over the world. Passive house standards are used as a basis for zero-energy architectures as it promotes energy optimization by design, construction and ventilation [32]. Discussion forums within the group offer a platform to share ideas on improvements, innovations, and real-world examples of buildings that meet Passive House or net-zero energy standards. Passive House forums such as the International Passive House Conference or other regional conferences can act as a perfect point for liaising with specialists and identifying how Passive House can be implemented in the further development of ZEB [48].

The *International Energy Agency* (IEA) is an international organization that focuses on cooperation in energy related research and policy analysis. The IEA Technology Collaboration Program on Energy in Buildings and Communities (IEA EBC) is a major area within the IEA that covers energy-efficient building solutions, including that of zero-energy buildings [49]. Networking can be provided at the annual meetings of the IEA EBC, workshops and conferences during which industry professionals can obtain information about the most recent achievements in provision of energy efficient buildings and measures [50]. Accordingly, the IEA EBC promotes continuous professional awareness of emerging trends in zero energy building design, making it especially pertinent to practitioners in sustainable architecture [49].

The *Net Zero Energy Coalition* (NZEC) is a non-profit peer-to-peer organization based in the U.S. It is an alliance of key industry stakeholders including architects, engineers, developers, policy professionals, and other practitioners who are pushing the development and use of zero-energy buildings. The alliance offers support to practitioners who want to create and develop zero-energy structures through best practice papers, case, and learning materials [51]. NZEC also provides opportunities for workshops, conferences and events to enable the professionals in the activities to share and interact. It is a platform to listen to professionals, learn about the latest technologies and policies that define the growth of zero-energy buildings. When joining the NZEC community, the professional discovers a pool of professionals with a noble goal of achieving the NZEC objectives of designing energy efficient architectural structures [52].

11.8 Essential Tools for ZEB Design

ZEB design requires a comprehensive approach that integrates cutting-edge simulation tools, established design standards, and real-world case studies. Here below are discussed tools and resources that can assist architects, engineers, and sustainability professionals in creating energy-efficient and environmentally responsible buildings. These tools encompass simulation software that navigate professionals through the complexities of ZEB design.

Energy modeling and simulation tools are essential in the design and evaluation of zero-energy buildings. They allow professionals to predict a building energy performance, assess various design strategies, and optimize systems for maximum efficiency before construction begins. Some of the most widely used packages that are used in assessing building energy performance and supporting sustainable design decisions are listed in Table 11.5.

Referring to Table 11.5 it is observed that there is a wide variety of software tools available to support building energy modeling and environmental analysis. Freeware license tools Energy Plus and HEED provide powerful simulation capabilities for whole-building energy use and residential design. Commercial software such as DesignBuilder and IES VE offer more integrated, user-friendly interfaces with added features like 3D modeling and detailed performance visualization. Most of these tools also complement each other through plugins or cloud-based platforms.

Energy Plus is a freeware license offering modeling and simulation of energy utilization, incorporation of renewable power systems, and a designing of heating, ventilation as well as air conditioning systems [55]. Thanks to software modularity, various aspects of energy flow in buildings can be analyzed by enabling users to develop energy simulations models with the actual operation along building design [69]. The integration of EnergyPlus with machine learning algorithms has been studied for refining the energy deferral projection, proves that such tools can be adjusted as per the contemporary building design [70].

TRNSYS, (Transient System Simulation Program), is one of the most effective simulation tools for modeling the energy system in building [65]. The capability of creating real load profiles makes it particularly useful for modeling actual renewable energy systems, energy storage systems, and dynamic building performance. Familiarizing with TRNSYS program will allow those professionals to design better balanced buildings to increase energy efficiency. For instance, Zhang and Hong equally stressed the applicability of TRNSYS in simulating HVAC systems and their energy efficiency determination [71]. EnergyPlus as well as TRNSYS has been found to be of high accuracy by various studies conducted to assess accuracy of building energy simulation software. For instance, the compatibility of TRNSYS with other simulation models has been examined to improve energy performance predictions [72].

Table 11.5 Popular building energy modeling and simulation tools

Software	Description	Developer	License	Source
EnergyPlus	Advanced simulation engine for whole-building energy analysis, including HVAC, lighting, and renewables	U.S. Department of Energy (DOE)	Freeware	[53–55]
OpenStudio	Interface and toolkit built on EnergyPlus for easier model creation and analysis	National Renewable Energy Lab (NREL)	Freeware	[56]
eQUEST	Quick and user-friendly tool for energy use and cost analysis based on DOE-2 engine	James J. Hirsch & Associates	Freeware	[57, 58]
RETScreen	Clean energy management software for feasibility, performance analysis, and financial modeling	Natural Resources Canada	Freeware	[59]
HEED (Home Energy Efficient Design)	Simple residential energy modeling tool for architects and designers	University of California, Los Angeles, and U.S. DOE	Freeware	[60]
Honeybee/ Ladybug (Plugins for Rhino/ Grasshopper)	Environmental analysis plugins integrating with EnergyPlus and Radiance	Mostapha Sadeghipour Roudsari	Freeware	[61]
DesignBuilder	3D modeling interface for EnergyPlus, offering dynamic thermal simulation and daylighting analysis	DesignBuilder Software Ltd	Commercial	[62]
IES VE (Virtual Environment)	Integrated performance analysis for building energy, daylight, airflow, and carbon modeling	Integrated Environmental Solutions (IES)	Commercial	[63]
TRNSYS	Transient system simulation tool ideal for complex energy systems, including renewables	University of Wisconsin & TESS	Commercial	[54, 64, 65]
SimScale	Cloud-based platform for HVAC, airflow, and thermal comfort simulations	SimScale GmbH	Commercial	[66]
Rhino + Grasshopper	3D modeling and parametric design software that integrates with Honeybee/ Ladybug plugins	Robert McNeel & Associates	Commercial	[67, 68]

11.9 Conclusion

The integration of energy efficient design technologies, renewable energy, and simulation, make ZEBs define new standards for energy conservation and sustainability in the built environment. But the process of transition does not come easy. Transitioning from conventional to zero energy buildings comes with financial challenges, and the growing need for cross disciplinary collaboration, including a development of new educational programs.

This complexity drives innovation. It is applied to many real-world renewable energy problems and shows how different approaches to energy simulation can be combined. The rapid development of wind, solar, and biomass, along with the importance of industry networks, education, and professional training, shows that this cannot happen without working together—it cannot be achieved by one organization alone.

ZEBs are not only architectural achievements, but a vision of the future that is energy self-sufficient and environmentally friendly. As we improve methods and use new technologies, we can ask ourselves: Is the difference only in the construction of buildings or also in the creation of a new way of life? This thought invites not only architects and engineers, but also the whole society, to engage in the possibilities that ZEBs offer.

References

1. Torcellini P, Pless S, Deru M, Crawley D (2006) Zero energy buildings: a critical look at the definition. In: ACEEE summer study pacific grove.
2. Harvey LDD (2013) Recent advances in sustainable buildings: review of the energy and cost performance of the state-of-the-art best practices from around the world. Annu Rev Environ Resour 38. https://doi.org/10.1146/annurev-environ-070312-101940
3. Singh P, Verma R (2010) Zero-energy buildings-a review
4. Austin MC, Chung-Camargo K, Mora D (2021) Review of zero energy building concept-definition and developments in latin america: a framework definition for application in Panama. https://doi.org/10.3390/en14185647
5. Yu Z, Gao C, Yang J, Wu J, Zhang H (2024) Overview of research and development of nearly zero energy buildings in China. Natl Sci Open 3(3). https://doi.org/10.1360/nso/20230083
6. Bentzen J, Engsted T (2001) A revival of the autoregressive distributed lag model in estimating energy demand relationships. Energy 26(1). https://doi.org/10.1016/S0360-5442(00)00052-9
7. Hawks MA, Cho S (2024) Review and analysis of current solutions and trends for zero energy building (ZEB) thermal systems. Renew Sustain Energy Rev 189:114028. https://doi.org/10.1016/J.RSER.2023.114028
8. Chen X, Yang H, Zhang W (2018) Simulation-based approach to optimize passively designed buildings: a case study on a typical architectural form in hot and humid climates. Renew Sustain Energy Rev 82:1712–1725. https://doi.org/10.1016/J.RSER.2017.06.018
9. Attia S (2011) State of the art of existing early design simulation tools for net zero energy buildings: a comparison of ten tools. Leed Ap (March)
10. Swan LG, Ugursal VI (2009) Modeling of end-use energy consumption in the residential sector: a review of modeling techniques. https://doi.org/10.1016/j.rser.2008.09.033

11. Hirst E, Lin W, Cope J (1977) A residential energy use model sensitive to demographic, economic, and technological factors. Q Rev Econ Financ 17
12. Haas R, Schipper L (1998) Residential energy demand in OECD-countries and the role of irreversible efficiency improvements. Energy Econ 20(4). https://doi.org/10.1016/S0140-9883(98)00003-6
13. Summerfield AJ, Lowe RJ, Oreszczyn T (2010) Two models for benchmarking UK domestic delivered energy. Build Res Inf 38(1). https://doi.org/10.1080/09613210903399025
14. Nesbakken R (1999) Price sensitivity of residential energy consumption in Norway. Energy Econ 21(6). https://doi.org/10.1016/S0140-9883(99)00022-5
15. Lee J, Lee M (2021) Educational effectiveness of a virtual design studio on zero-energy building designs: the case of the U.S. solar decathlon design challenge. J Green Build 16(4):249–262. https://doi.org/10.3992/jgb.16.4.249
16. Ozturk HK, Canyurt OE, Hepbasli A, Utlu Z (2004) Residential-commercial energy input estimation based on genetic algorithm (GA) approaches: an application of Turkey. Energy Build 6(2). https://doi.org/10.1016/j.enbuild.2003.11.001
17. Saha GP, Stephenson J (1980) A model of residential energy use in New Zealand. Energy 5(2). https://doi.org/10.1016/0360-5442(80)90005-5
18. Ascanio-Villabona JG, Tarazona-Romero BE, Duran-Sarmiento MA, Lengerke O, Betancur-Arboleda LA (2023) Evaluation of the thermal performance of housing envelopes as passive cooling systems. Sustain Eng Innov 5(2):177–188. https://doi.org/10.37868/SEI.V5I2.ID227
19. Li W et al (2017) Modeling urban building energy use: a review of modeling approaches and procedures. https://doi.org/10.1016/j.energy.2017.11.071
20. Fumo N, Rafe Biswas MA (2015) Regression analysis for prediction of residential energy consumption. https://doi.org/10.1016/j.rser.2015.03.035
21. Crawley DB, Hand JW, Kummert M, Griffith BT (2008) Contrasting the capabilities of building energy performance simulation programs. Build Environ 43(4). https://doi.org/10.1016/j.buildenv.2006.10.027
22. Fabbri R, Gabrielli L, Ruggeri AG (2018) Interactions between restoration and financial analysis: the case of Cuneo War Wounded House. J Cult Herit Manag Sustain Dev 8(2). https://doi.org/10.1108/JCHMSD-05-2017-0026
23. Khan HS, Asif M, Mohammed MA (2017) Case study of a nearly zero energy building in Italian climatic conditions. Infrastructures (Basel) 2(4). https://doi.org/10.3390/infrastructures2040019
24. Musall E et al (2016) Net zero energy solar buildings: an overview and analysis on worldwide building projects. https://doi.org/10.18086/eurosun.2010.06.16
25. Xu X, Feng G, Chi D, Liu M, Dou B (2018) Optimization of performance parameter design and energy use prediction for nearly zero energy buildings. Energies (Basel) 11(12). https://doi.org/10.3390/en11123252
26. Slimani J, Kadrani A, EL Harraki I, and hadj Ezzahid E (2022) Long-term bottom-up modeling of renewable energy development in Morocco. Adv Sci Technol Eng Syst J 7(5). https://doi.org/10.25046/aj070515
27. Heiker M, Kraume M, Mertins A, Wawer T, Rosenberger S (2021) Biogas plants in renewable energy systems—a systematic review of modeling approaches of biogas production. https://doi.org/10.3390/app11083361
28. ALRubeei IRN, Idi SN, Hussein Alsammak IL, AlRikabi HT, Mutar HA, Alaidi AHM (2025) Using artificial intelligence for enhancement of solar cell efficiency in the of Iraq. Herit Sustain Dev 7(1):197–210. https://doi.org/10.37868/HSD.V7I1.1067
29. Clarke L, Gleeson C, Winch C (2017) What kind of expertise is needed for low energy construction? Constr Manag Econ 35(3). https://doi.org/10.1080/01446193.2016.1248988
30. Phius training update: in-person courses, opportunity in massachusetts, and more|Phius Phius Training Update: In-Person Courses, Opportunity in Massachusetts, and More. Accessed 07 Jan 2025. [Online]. Available: https://www.phius.org/phius-training-update-person-courses-opportunity-massachusetts-and-more

31. Rabaia MKH et al (2021) Environmental impacts of solar energy systems: a review. Sci Total Environ 754. https://doi.org/10.1016/j.scitotenv.2020.141989
32. Lacey-Barnacle M, Smith A, Foxon TJ (2023) Community wealth building in an age of just transitions: exploring civil society approaches to net zero and future research synergies. Energy Policy 172. https://doi.org/10.1016/j.enpol.2022.113277
33. Lee J, Lee M (2022) Curriculum development for an integrated zero-energy building design-in case of the U.S. Department of Energy Solar Decathlon Design Challenge. J Arch Inst Korea 38(10). https://doi.org/10.5659/JAIK.2022.38.10.131
34. Mohammadi A, Mousavi SM (2022) Feasibility study of net-zero energy residential buildings in hot and humid climates: a case study of Iran. J Teknol 84(2). https://doi.org/10.11113/jurnalteknologi.v84.17662
35. Online education: the benefits of professional study—global energy certification. Accessed 07 Jan 2025. [Online]. Available: https://globalenergycertification.org/online-education-the-benefits-of-professional-study/
36. Online education—available anytime—zero energy project. Accessed 07 Jan 2025. [Online]. Available: https://zeroenergyproject.com/online-education-available-online-anytime/
37. Training programs—midwest renewable energy association. Accessed 07 Jan 2025. [Online]. Available: https://midwestrenew.org/training-programs/
38. Explore tools, trainings, and resources|better buildings initiative. Accessed 07 Jan 2025. [Online]. Available: https://betterbuildingssolutioncenter.energy.gov/better-plants/tools-and-trainings
39. Abdellah RH, Masrom MAN, Chen GK, Mohamed S, Manap N (2018) An investigation on current green design practices towards net-zero energy buildings (NZEBs) application in Malaysia healthcare projects. J Technol Manag Bus 5(3). https://doi.org/10.30880/JTMB.2018.05.03.007
40. Cohen R, Desai K, Elias J, Twinn R (2021) Net zero carbon: Energy performance targets for offices. Build Serv Eng Res Technol 42(3). https://doi.org/10.1177/0143624421991470
41. View of net zero healthy building: defining the terms and criteria with bibliometric analysis. Accessed 07 Jan 2025. [Online]. Available: https://jurnalsaintek.uinsa.ac.id/index.php/EIJA/article/view/1419/808
42. Leadership C, Document D (2010) Canada green building council. In: LEED Canada for new construction and major renovations
43. Council WGB (2024) Advancing net zero
44. Vosoughkhosravi S, Dixon-Grasso L, Jafari A (2022) The impact of LEED certification on energy performance and occupant satisfaction: a case study of residential college buildings. J Build Eng 59. https://doi.org/10.1016/j.jobe.2022.105097
45. Cao X, Dai X, Liu J (2016) Building energy-consumption status worldwide and the state-of-the-art technologies for zero-energy buildings during the past decade. Energy Build 128. https://doi.org/10.1016/j.enbuild.2016.06.089
46. Phius|Phius Zero is the goal. Phius is the means. Accessed 07 Jan 2025. [Online]. Available: https://www.phius.org/
47. Omrani S, Garcia-Hansen V, Capra BR, Drogemuller R (2017) On the effect of provision of balconies on natural ventilation and thermal comfort in high-rise residential buildings. Build Environ 123. https://doi.org/10.1016/j.buildenv.2017.07.016
48. Su B (2008) Building passive design and housing energy efficiency. Archit Sci Rev 51(3). https://doi.org/10.3763/asre.2008.5133
49. International energy agency's energy in buildings and communities programme||IEA EBC. Accessed 07 Jan 2025. [Online]. Available: https://www.iea-ebc.org/
50. Jaysawal RK, Chakraborty S, Elangovan D, Padmanaban S (2022) Concept of net zero energy buildings (NZEB)—a literature review. https://doi.org/10.1016/j.clet.2022.100582
51. Urge-Vorsatz D et al (2020) Advances toward a net-zero global building sector. https://doi.org/10.1146/annurev-environ-012420-045843
52. Net-zero energy buildings: a path to sustainable living—Engineering Heritage Journal (GWK). Accessed 07 Jan 2025. [Online]. Available: https://enggheritage.com/gwk-02-2024-81-87/

53. EnergyPlus. Accessed 19 June 2025. [Online]. Available: https://energyplus.net/
54. Ascanio-Villabona J, Orejarena-Osorio N, Castillo-Leon NY, Tarazona-Romero BE, Jaimes-Quintero KT (2024) Thermal comfort evaluation using different glazing technologies in an energy model of rural housing in a cold climate. Herit Sustain Dev 6(2):735–748. https://doi.org/10.37868/HSD.V6I2.922
55. EnergyPlus|Department of Energy. Accessed 07 Jan 2025. [Online]. Available: https://www.energy.gov/eere/buildings/articles/energyplus
56. OpenStudio Coalition User Docs. Accessed 19 June 2025. [Online]. Available: https://openstudiocoalition.org/
57. eQUEST LEED Training|Energy-Models.com. Accessed 19 June 2025. [Online]. Available: https://energy-models.com/training/equest-leed-case-study
58. eQUEST. Accessed 19 June 2025. [Online]. Available: https://doe2.com/equest/
59. RETScreen—Natural Resources Canada. Accessed 19 June 2025. [Online]. Available: https://natural-resources.canada.ca/maps-tools-publications/tools-applications/retscreen
60. HEED: Home Energy Efficient Design|Society of Building Science Educators. Accessed 19 June 2025. [Online]. Available: https://www.sbse.org/resources/heed
61. Ladybug Tools|Home Page. Accessed 19 June 2025. [Online]. Available: https://www.ladybug.tools/
62. DesignBuilder Software Ltd—Home. Accessed 19 June 2025. [Online]. Available: https://designbuilder.co.uk/
63. Integrated Environmental Solutions|IES. Accessed 19 June 2025. [Online]. Available: https://www.iesve.com/
64. Welcome|TRNSYS : Transient System Simulation Tool. Accessed 19 June 2025. [Online]. Available: https://www.trnsys.com/
65. Klein SA, Beckman WA (2007) TRNSYS 16: a transient system simulation program: mathematical reference. Trnsys 5
66. Simulation Software|Engineering AI in the Cloud|SimScale. Accessed 19 June 2025. [Online]. Available: https://www.simscale.com/
67. Grasshopper—algorithmic modeling for Rhino. Accessed 19 June 2025. [Online]. Available: https://www.grasshopper3d.com/
68. Rhino—Rhinoceros 3D. Accessed 19 June 2025. [Online]. Available: https://www.rhino3d.com/
69. Dols WS, Emmerich SJ, Polidoro BJ (2016) Coupling the multizone airflow and contaminant transport software CONTAM with EnergyPlus using co-simulation. Build Simul 9(4). https://doi.org/10.1007/s12273-016-0279-2
70. Pensado-Mariño M, Febrero-Garrido L, Eguía-Oller P, Granada-álvarez E (2021) Feasibility of different weather data sources applied to building indoor temperature estimation using LSTM neural networks. Sustainability (Switzerland) 13(24). https://doi.org/10.3390/su132413735
71. Kuznik F, Virgone J, Johannes K (2010) Development and validation of a new TRNSYS type for the simulation of external building walls containing PCM. Energy Build 42(7). https://doi.org/10.1016/j.enbuild.2010.01.012
72. Nayak AK, Hagishima A (2020) Modification of building energy simulation tool TRNSYS for modelling nonlinear heat and moisture transfer phenomena by TRNSYS/MATLAB integration. In: E3S web of conferences. https://doi.org/10.1051/e3sconf/202017225009

MIX
Papier aus verantwortungsvollen Quellen
Paper from responsible sources
FSC® C105338

If you have any concerns about our products,
you can contact us on
ProductSafety@springernature.com

In case Publisher is established outside the EU,
the EU authorized representative is:
Springer Nature Customer Service Center GmbH
Europaplatz 3, 69115 Heidelberg, Germany

Printed by Libri Plureos GmbH
in Hamburg, Germany